SUSTAINABLE POTATO PRODUCTION IN HILLY REGION OF INDIA

SUSTAINABLE POTATO PRODUCTION IN HILLY REGION OF INDIA

UMA SAH
K.D. KOKATE
B.P. SINGH
V.K. GUPTA
S.K. DUBEY

BIOTECH
2014
BIOTECH BOOKS
DELHI

Edition : 2014

ISBN : 978-81-7622-296-9

Published by : **BIOTECH BOOKS**
4762-63/23, Ansari Road, Darya Ganj,
New Delhi - 110 002
Phone : 23262132
email : biotechbooks@yahoo.co.in

PRINTED IN INDIA

Preface

The hilly region particularly North Eastern (NE) states comprises of seven states namely Arunachal Pradesh, Assam, Manipur, Meghalya, Mizoram, Nagaland, Sikkim and Tripura. These states lies between 21.5° and 29.5° North latitude and 85.5°– 97.5° East longitudes and represents distinct – agro climatic area of the country. These states have a geographical area of 25.08 million hectares (7.79% of the country) with human population of 38.495 million (3.75% of India). The population density varies from 10 persons per square km in Arunachal Pradesh to 323 persons per square km in Tripura. This affects the production system and the cropping intensity which is lowest in Arunachal Pradesh and highest in Tripura respectively. North eastern region is highly diverse in climatic conditions, which varies from place to place and even within a short distance huge climatic diversity can be observed. The latitudinal location and its elevation from the sea level considerably influence the climate which varied from alpine in Arunachal Pradesh and Sikkim to sub-tropical in Assam and Tripura.

The climatic conditions are highly suitable for the cultivation of large number of horticultural crops including fruits and vegetables indicating the vast potential for their further improvement. *Jhumming* is the dominant system of cultivation in the region, and is regarded to be the first step in the transition from food gathering or hunting to food production. It is primitive mode of agriculture evolved as a reflex action under the stress of peculiar geophysical and ecological situations of the hills of North eastern region.

The present book therefore attempts to highlight some of the major issues pertaining to potato cultivation in the North Eastern Hilly (NEH) region. These issues include – problems in potato cultivation, issues related to quality seed potato production, the potential ways and means to overcome the prevailing production bottlenecks. At the same time, book also presents some of the 'experiences and impact of on-farm experimentations in northern hills of India.

The book contains 30 chapters which has been divided into five parts. Part-1 deals with the Agrarian Scenario in North Eastern Region which include the traditional potato production scenario and the related aspects. Part-2 discusses Research Backstopping for Potato Production in NEH region. Part-3 of the book highlights Technological Backstopping for Potato Production in the region and Part-4 emphasized on Transferrable Technologies and on-farm adaptive researches for Enhanced Potato Production in NEH. Last part of the book focuses on the on-farm experimentations, entrepreneurship promotion and impact of extensioon programmes related to potato technologies in the northern hills of India region. Based on the experiences of the findings, some of the meaningful conclusions for sustainable potato R&D for this region have been suggested.

Authors are confident that this book will act as the ready reference to the researches, extension personnel, students, NGOs and all related stakeholders who are involved in potato R&D in North eastern region.

Authors

Contents

PART - 1

AGRARIAN SCENARIO IN NORTH EASTERN REGION

CHAPTER 1

Agricultural Situations in North Eastern (NE) Region of India

The North eastern region comprising of states of Arunachal Pradesh, Assam, Manipur, Meghalaya, Mizoram, Nagaland, Sikkim and Tripura lying between 21.5° and 29.5° North latitude and 85.5 ° - 97.5 ° East longitudes, represents distinct-agro climatic area of the country. These hill states have a geographical area of 25.08 million hectares (7.79% of the country) and have a population of 38.495 million (3.75% of India). The population density varies from 10 persons per square km in Arunachal Pradesh to 323 persons per square km in Tripura. This affects the production system and the cropping intensity resulting in the lowest production and cropping intensity in Arunachal Pradesh and highest in Tripura. North eastern region is highly diverse in climatic conditions, which varies from place to place and even within a short distance huge climatic diversity can be observed. The latitudinal location and its elevation from the sea level considerably influence the climate which varies from alpine in Arunachal Pradesh and Sikkim to sub-tropical in Assam and Tripura.

The temperature of North Eastern region varies due to change in altitude. It reaches maximum of 27 to 29 °C during June-August and a minimum of 5-8°C during December-January. Wide spread low clouds and high humidity maintains uniform temperature and the average temperature ranges between 20-29 °C. The relative humidity remains quite high (85% or more) during monsoon season. It rarely is below 75% which is very congenial for the development of various diseases in different crops including potato.

The climatic conditions are highly suitable for the cultivation of large number of horticultural crops including fruits and vegetables indicating the vast potential for their further improvement and also for the development of horticultural industry in the region. Traditionally shifting cultivation, known as *Jhumming* is the dominant system of cultivation in the region, and is regarded to be the first step in the transition from food gathering or hunting to food production. It is primitive mode of agriculture evolved as a reflex action under the stress of peculiar geophysical and ecological situations of the hills of North eastern region.

The vast area of hills interspersed with fertile valleys represents the agro-climate of unique diversity falling within altitude range of 15m to 1500 m of average mean sea level and above. Coupled with this is the abundance of natural resources including land, water and vegetation with well distributed precipitation and deep alluvial loamy soils having pH ranging from 4.5 to 6.5, which allows to growing of wide variety of agricultural/ horticultural crops within a short distance. The region experiences high rainfall from 200 to 455 cm, a major part (70 to 90%) is received during the pre-monsoon and monsoon period (April to October). Although, it is a high rainfall area, but there could be periods of scanty rainfall during south east monsoon season. In all, North Eastern region receives 40 million hectare meters of precipitation which is about 21% of country's total precipitation.

Historically, agriculture has remained underdeveloped in the region as a result of proportionately less cultivation area, poor irrigation systems, low or no adoption of improved production technologies including application of in organics and pesticides, coupled with general remoteness of the region and land tenure systems.

Potato is being cultivated in the region traditionally and a large number of European varieties are still cultivated in pockets. Some of the popular old introductions are; *Up-to date, President, Great Scot, Royal Kidney, Red Pimpernell* which are poor yielder and highly susceptible to late blight. Recently however, some of the promising high yielding and late blight resistant varieties like Kufri Jyoti, Kufri Khasigaro, Kufri

Megha, Kufri Giriraj, Kufri Shailja and recently Kufri Himalini and Kufri Girdhari have been developed and introduced for commercial cultivation.

Almost 10 percent of the geographical area of the country and about 4.5 percent of its population comes under northeastern (NE) region of India. Also, about 10 percent of the country's total potato area lies in this region. The area under potato as a percentage of the net-cropped area is about four times of the national level. However per unit production of potato in all other NE states except Tripura (18.52 t/ha) has been all time low (4-8 t/ha) as compared to national average yield of 19 t/ha. The reasons attributing to low potato yields are many. These include inadequate availability of crop inputs like healthy seed vis-à-vis poor management practices being followed by the potato growers and also prevalence of optimum temperature and high humidity during crop season in the region favouring perpetuation of serious diseases like late blight, brown rot/bacterial wilt, etc. Importantly, the production of breeder seed in this region is not possible owing to several biological and environmental factors. Prominent among them are prevalence of virus vectors throughout the crop stages followed by late blight and bacterial wilt which are the major reasons in the non-production of quality seed potato. The breeder seed, therefore, is transported to the different NE states from other parts of the country, due to which it is oftenly not available in the right physiological age. The available seed hence in the region is highly degenerated and is mainly responsible for low yields. Table in Annexure 4 shows relative status of area, production and yield of potato in different states of NE region which shows the significance of potato in this region.

Prevailing Potato Based Cropping Sequences in the Region

Every state in this region has its own potato based crop rotation. Prominent among them are with vegetables and paddy. Some of the major crop rotations (state wise) are mentioned below :

- **Arunachal Pradesh :** Paddy-potato, maize-potato, small millets-potato, groundnut–potato.

- **Assam** : Paddy (*Ahu*)-paddy (*sali)*-potato
- **Manipur** : Potato-potato, maize-potato, paddy-potato.
- **Meghalaya** : Potato-potato, potato-cauliflower, potato-cabbage, potato-radish, potato-paddy.
- **Mizoram** : Paddy-potato, pulses-potato.
- **Tripura** : Paddy-potato, vegetables-potato.

Agro-eco-situations Prevailing in the Region

The details of the production situations for potato crop prevailing in all the states in north eastern region have been shown in Annexure I. The table shows that potato in this region could be grown in plains (up to 500 m from msl) to as high as above 2400 m from msl. Therefore, production technologies and the production situation characteristics need to be matched.

Status of Plant Nutrients and Chemicals Consumption in NE Region

Consumption of Plant Nutrient in NE Region (2003-2004) Kg/ha

Table 1.1 : Consumption of plant nutrient in NE Region (2003-2004) (Kg/ha)

State	*N*	*P*	*K*	*Total*
Assam	19.18	10.56	9.07	38.81
Arunachal	1.56	0.90	0.42	2.88
Manipur	86.54	12.00	6.40	104.94
Meghalaya	10.52	6.16	0.48	17.16
Mizoram	4.94	5.16	3.62	13.72
Nagaland	1.12	0.85	0.16	2.13
Sikkim	5.05	3.49	1.19	9.73
Tripura	16.55	8.63	5.28	30.46
NE mean	18.85	9.07	7.17	35.09
India Mean	58.72	22.75	8.66	90.12

(Agril. statistics at a Glance, 2005; DOAC, Govt. of India)

The information as furnished in Table 1.1 clearly indicates that the application of inorganic fertilizers in the region is very low and whatever the fertilizers are applied are generally applied in the valley areas or the irrigated eco-systems while upland eco-systems are mainly organic based. Nagaland consumed the lowest quantity of inorganic fertilizers followed by Arunachal Pradesh while Manipur applies the highest doses of inorganic fertilizers. After the introduction of organic production system concept, Mizoram and Sikkim have been declared as the organic states where the fertilizer consumption shall further decrease.

Consumption of Plant Protection Chemicals in NE States

The information presented in Table 1.2 indicates the extent of application of chemical pesticides for the management of various diseases and insect pests. The pesticide consumption indicated that these are applied only to vegetables or the high value crops in the valley areas. In shifting cultivation areas neither the fertilizers nor the pesticides are applied.

Table 1.2 : Consumption of plant protection chemicals in NE States (Tons / year technical grade)

State	*95-96*	*97-98*	*99-2000*
Assam	316.00	284.00	260.00
Arunachal	22.00	18.00	17.00
Manipur	41.00	20.00	21.00
Meghalaya	20.00	8.00	8.00
Mizoram	21.00	17.00	19.00
Nagaland	9.00	9.00	10.00
Sikkim	26.00	16.00	0.16
Tripura	25.00	19.00	17.00
Total NE	480.00	391.00	352.00

Keeping the low productivity and the potential of further improvement in the region Govt. of India, Ministry of Agriculture and Co-operation (Horticulture Division) initiated the Central Sector Scheme in the form of "Technology Mission for Integrated development of Horticulture in North Eastern

Region including Sikkim, J&K, HP and Uttarakhand" and Central Potato Research Institute and its regional station at Shillong have identified certain constraints in low potato production as mentioned below and those need to be addressed through appropriate research and extension interventions.

Constraints in Potato Production in North Eastern India

Non-Availability of Healthy Seed

Low productivity of potato in North eastern region is mainly due to non-availability of healthy seed. Moreover, being a vegetatively propagated crop, 25-30 q seed is required for a hectare resulting in about 50% of the total cost of cultivation on seed itself. The region being situated far from northern seed producing areas can not afford to procure healthy seed potato from north and farmers are compelled to use locally available degenerated seed year after year resulting in low yields. Further, the breeders' seed supplied to different NEH states from North India is not available in the right physiological stage at the time of planting because the sprouts become very lanky due to temperature variation.

Inadequate and Untimely Availability of Essential Inputs

Most of the farmers in the region are not able to purchase required quantity of fertilizers and pesticides. It is known that potato is a high input crop and if balanced doses of fertilizers are not applied, one cannot have a good harvest.

Prevalence of Destructive Diseases

Serious diseases like late blight and bacterial wilt/brown rot are prevalent in the region which is responsible for low productivity. Though, suitable management practices *viz.* late blight resistant varieties, management of late blight through chemicals like Dithane M-45, Ridomil MZ and brown rot management through stable bleaching powder are available, farmers are either not applying the chemicals or even if applied the recommended doses are not followed. Secondly, farmers are not getting the required seed material of late blight resistant varieties like Kufri Giriraj, Kufri Megha, Kufri Kanchan, Kufri Girdhari and even Kufri Jyoti.

Lack of Knowledge about Improved Potato Production Technologies

During formal and informal interface with farmers', it was observed that they were totally unaware/ignorant about the improved production technologies such as preparation of field, disease free seed, spacing, fertilizer application, management of diseases/pests and suitable potato varieties for the area as developed and recommended by CPRI. Even the Government. Farms do not adopt the improved technologies.

Problem of Language

Language of each state is different and is a great barrier in disseminating the knowledge to the farmers. Trying local person either from the Agriculture/Horticulture Department or from the ICAR Stations as interpreter,though,helped a lot but even then its impact is not well known.

Jhum or Shifting Cultivation

Jhuming or shifting cultivation is a common practice among the tribals of Meghalaya, Nagaland, Arunachal Pradesh and Mizoram. This has degraded the land resulting in soil erosion and loss of nutrients and is mainly responsible for low yields in the region. There is no organized settlement policy for tribal people.

Organic Farming

Recently, some of the states of the region like Mizoram and Sikkim have declared their states as organic states and others are also following the suit. Not much work has been done on organic farming in the field of potato production. The present day potato varieties require high input of fertilizers and these directly affect the yield.Hence, potato cultivation solely on basis of organic farming does not seem to have much scope.

Marketing Problems

Bad road conditions, price fluctuation, involvement of middleman and lack of processing industry are some of the major marketing related constraints in this region.

Storage Related Problems

High temperature and high humidity during storage period, lack of cold storage facility near the production sites and lack of proper scientific methods of storage under ambient condition are some of the major storage constraints.

Besides, fixation of applied phosphorous due to high aluminum content in the acidic soils of the region, resulting in poor yield; high rainfall in the main crop season (March-July) resulting in infestation of weeds and attack of many diseases and pests, lack of knowledge of disease free seed production and appropriate storage practices and inappropriate methods of planting *viz.* in heap, in pits and broadcasting are also affecting the potato yield in the area.

Ways and Means to Overcome the Constraints

Strengthening the Supply of Quality Seed and its Timely Availability at Planting

Usually, the breeder seed is supplied to these NE hilly (NEH) states from Kufri, Shimla during November-December. It is planted in different states according to altitudes during February-April each year. The seed is exposed to variable temperatures from Himachal Pradesh to NEH region right from transportation up to utilization. Storage is a basic problem, hence, the seed potato is stored at ambient temperatures resulting in early sprouting and sprouts become very lanky by the time of planting. Arrangement may be made in such a way that breeder's seed reaches there by mid of February. The quantity of the breeder's seed should also be increased so as to meet the full demand of different NEH states.

Laying on-farm demonstration trials in different states by the state department of Agriculture/Horticulture using improved production technology

The scientists from potato research station at Shillong, Meghalaya are trying their best to train the farmers in production and protection technologies. It is well known that 'seeing is believing', hence on-farm trials may be arranged in representative villages of different states using improved potato technologies versus traditional methods. The comparison of both the methods will certainly attract the farmers towards improved technologies. The Krishi vigyan Kendras (KVKs) in the respective districts in the region have significant role to play for technology assessment and refinement to make them situation appropriate.

Identification of Areas Suitable for Quality Seed Production

It is important to make this region self dependent in seed potato which may be available locally or from the nearby places located at a reasonable distance. Manipur, Arunachal Pradesh and Sikkim are having their farms at higher elevations which are suitable for quality seed production. If the quality seed production is strengthened in these states, seed supply can be managed to other sister states also. Alternatively, other states may also select areas for quality seed production in their own states.

Improvisation of Storage Structures/Methods

The locally available pit/heap storage methods need to be improved a lot. The bamboo support as recommended in pit storage for North-Indian plains may also hold good for this region. There is enough bamboo available in the region, therefore this modification will be easy, to follow. For seed storage, diffused light stores will be promising. For ware potatoes, storage in heaps with proper aeration facility and thick cover of locally available straw/hay would be more desirable. In lower areas, Govt. or local bodies may take initiative to construct cold storage facility for both seed storage (*i.e.* 2-4°C) and table and processing potatoes (*i.e.* 10 to 12°C with sprout suppressant application).

Development of Technology for Organic Farming

Suitable agronomic package of practices for organic potato farming may be developed keeping in view the requirement of states which are shifting to organic farming and imposition restrictions/ban on the use of chemical fertilizers and plant protection chemicals. The potatao research center at Meghalaya has also popularized vermicomposting in the state. The experiences of this technology may be replicated in other states as well.

Regulating the Seed Supply

Potato seed produced in the high hills during summer may be supplied regularly for planting in low elevation areas in winter/*rabi*. The state Government may be requested to seriously look into this aspect.

Publications of Extension Literature in Local Languages

Package of practices for potato production may be printed in local languages by the State Governments/ICAR institutes/ CAU and may be supplied to the farmers for its best use by them.

Marketing

Necessary policy initiatives in the form of minimum support price for the potato growers, provision of improved on-farm storage facility, provision of cold storage at the approachable distance and information support of marketing intelligence are some of the potential intervention in this regard.

Status of Horticultural Research in the Region

Research Infrastructure

The Indian Council of Agricultural research (ICAR) is carrying out horticulture research in the region through NEH Research Complex, Barapani (Meghalaya) and its regional stations; National Research Centre for Orchids, Pakyong (Sikkim); Central Potato Research Station, Upper Shillong (Meghalaya) and Central Plantation Crops Research Institute Regional Station, Kahikuchi (Assam). In addition, Assam Agricultural University, Jorhat and its research stations are contributing to horticulture research and development in Assam. Further, 11 research centres of All India Coordinated Research Projects on Vegetables, Potato, Tuber crops, Palms and Betelvine located at AAU, Jorhat, Tinsukia and Kahikuchi are conducting multilocational trials for identifying promising cultivars for the region. Twelve Krishi Vigyan Kendras (KVKs) in the region and one Trainers Training Centre (TTC) in Meghalaya are providing extension back-up and support towards popularization of improved technology and development of skilled manpower for various horticultural programmes.

Status of Research

Concerted research efforts have been made to identify a large number of improved varieties and production technologies of fruits, vegetables and tuber crops including potato and plantation crops suitable for the region.

A brief analysis of strength and weakness of the region has been brought out as below :

(i) Strength

- Region highly suitable for cultivation of large number of horticultural crops due to rich soil and agro-climatic conditions.
- Availability of wide genetic resource base and varying production systems in the region.
- Reasonable R&D infrastructure for technology generation and transfer.

(ii) Weakness

- Limited availability of location-specific improved cultivars of horticultural crops for the region.
- Low levels of productivity of horticultural crops.
- Limited availability of quality seed, planting material and other inputs.
- Limited awareness towards improved production practices, and
- Non availability of trained manpower.

Status of Potato Cultivation

Potato is an important vegetable crop of the region. The

CPRI regional station in Meghalaya has developed number of improved varieties and appropriate management practices for the crop. The productivity is fairly high particularly in Tripura (18.5 t/ha) and the state has achieved distinction in producing TPS on commercial scale. Kufri Megha and Kufri Jyoti have been recommended for main and autumn season crops for the region. Among the recently developed cultivars, Kufri Giriraj, resistant to late blight, is widely under cultivation. Besides, CPRI has recently developed Kufri Shailja and Kufri Himalini which are also resistant to late blight. The red skinned potato cultivar, Kufri Kanchan is very popular in Sikkim and some other parts of the region. A number of improved cultural practices have also been developed for the region.

Seed Potato Production Units in NE States

State	*Seed Potato units*	*Crops grown*
Arunachal Pradesh	Govt. Foundation Seed Multiplication Farm, Tawang	Potato, Cabbage, Cauliflower
Assam	No specific farm devoted to Potato	Potato along with other vegetables
Manipur	Regional Potato Farm, Mao	Potato
Meghalaya	Experimental & Research Station, Upper Shillong	Potato
Mizoram	No specific farm devoted to Potato	Potato and other vegetables
Nagaland	1. Tfutsero Res. Farm, Phek	Potato
	2. Aghanato Farm, Zunhepoto	Potato
Sikkim	1. Rabong Potato Seed Farm, Rabangla (S. Sikkim)	Potato
	2. Seed Potato Farm, Hilley (W. Skkim)	Potato
Tripura	Horticulture Res. Complex, Nagichhera	True Potato Seed, Vanilla

CHAPTER 2

Traditional Potato Cultivation in NE Region : A Case of Meghalaya

About Meghalaya

Meghalaya is one of the seven sister states of North Eastern Hill (NEH) Region of India. It came into inception as an independent and full fledged state in January 1972. The state shares its boundaries with Assam in the north and east and with Bangladesh in south and west. The state of Meghalaya appears to be on an imposing table land or plateau and is bordered by deep gorges and land sloping steeply towards the plains. The state occupies about 22,500 sq kms and consists predominantly of mountainous terrain with narrow valleys in-between and strips of plain land in southwest.

Meghalaya state is situated between 21° 71 and 26o 71and 26o10 N latitude and 89.47 o E longitude. The state is divided into seven districts viz. Jaintia hills, East Khasi Hills, West Khasi Hills, Ri- bhoi, East Garo hills, West Garo hills and South Garo hills district.

Climate

In general, Meghalaya has a monsoon climate with natural vegetation profile ranging from tropical to temperate type. Average rainfall is 11,000 mm per annum. The climate however varies with the altitude and physiographic difference of landmass. Most of the rainfall is received in the months from April to September. The temperature in the state varies from a minimum of 4.6oC to a maximum of 29.1 The relative humidity during the rainy season average around 85%.

Demography

The total population in the state is 23.57 Lakh and the population density is 103 persons per square kilometers. More than 80 percent of the population lives in rural areas and depends on agriculture as their mainstay. Majority of the population in the State belongs to the three ancient hill tribes namely, Khasi, Jaintias and Garos. The existing matrilineal as well as matriarchal social system among all of these hill tribes makes these societies unique in India. The literacy rate in the state is about 51 percent.

Land Use Pattern

About 207 thousand hectares of land is under cultivation in the state which is about 9.23 percent of the total geographical area of the state. Agricultural in this hilly state is mainly rainfed with only 47 thousand hectares of net irrigated area which is approximately 23 percent of the net cropped area. About 44 thousand hectares of land is sown more than once in a year in the state with the cropping intensity as low as 120 percent.

The land ownership pattern in this tribal state is unique. The agricultural land in the state is owned by certain clans or is community owned which is under control of the village headman or sometimes the land is owned by individual farmers, depending upon the locality. The landless farm families lease in the land for cultivation.

Shifting cultivation – commonly called as the slash and burn method wherein the cultivators clear the selected forest area from all vegetation including trees and use the land for cultivation for some years before they shift to another land for cultivation, is still in practice in the North Eastern Hill region including Meghalaya. In Meghalaya state the minimum area under shifting cultivation one time or the other is approximately 2650 sq kms and about 52,290 farm families are practicing shifting or in other words *Jhum* cultivation in the state with the *Jhum* cycle of 5-7 years. Annually about 530 sq kms of area is brought under shifting cultivation in the state (Annon, 2002).

The main crop grown in the state is Rice which is grown on half of the cultivated area (103 th.ha.) in the state. Potato is the second important crop in the state with a total of 20 th. ha. of land under its cultivation. Other important grown in the state are maize, oilseeds, pulses, fiber crops, small millets, tobacco, wheat are the important crops grown in the state (Table 2.1).

Table 2.1 : Area, production and productivity of important crops grown in Meghalaya state

Sl.No.	*Crop*	*Area (th.ha.)*	*Production (th. tonnes)*	*Productivity (kg/ ha.)*
1.	Rice	103	166	1611.7
2.	Potato	20	182	9100.0
3.	Maize	17	24	1411.8
4.	Oilseeds	9	6	667.0
5.	Pulses	5	4	8000.0
6.	Small millets	3	2	667.0
7.	Cotton lint	7	8	1142.8*
8.	Jute	4	32	1440.0
9.	Mesta	5	1	756.0
10.	Wheat	4	7	1750.0
11.	Tobacco	1	1	1000.0
12.	Chillies	2	1	500.0

Source : Basic Statistics of NER, 2005, Government of India.

• Production in thousand bales of 170 kgs each.

Potato in Meghalaya

Potato is one of the most important crops grown in the state especially in the hilly regions. Potato crop occupies a major niche in the cropping pattern as well as it significantly contributes to the rural agrarian economy prevailing in Meghalaya state. Potato is grown on over 18.1 thousand hectares of land which constitutes a significant percentage *i.e.*, 8.56 percent of the total cultivable area in this tribal hill state and this which is highest among all the potato producing states in India. However, the average productivity of potato in the state is 9.2 t/ha. which is far below the national average of 17t/ha. Several factors like rainfed conditions, non-availability of good quality seed, heavy disease incidence etc., attribute to this yield gap.

Potato is cultivated in Meghalaya primarily in two hill districts namely, East and West Khasi Hills districts. These two districts together contribute around 95 percent of the total land under the potato crop and 96 percent of the total potato production in the state (Table 2.2). These two hills districts, as the name reflects are inhabited by the Khasi hill tribe. Thus potato in Meghalaya is primarily cultivated by the khasi tribe.

Potato crop is grown in Meghalaya state in two seasons *i.e.*, summer and autumn season. The summer season is the main potato-growing season in the state that extends from the month of February to June-July. Autumn season on the other hand starts from the month of July-August and ends in the month of November-December. Autumn season is comparatively lesser utilised in the district for potato cultivation.

Table 2.2 : Area, Production and Yeild of potato in different districts of Meghalaya

Sl.No.	*District*	*Area(ha.)*	*Production (in MT)*	*Yeild (Kg/ha.)*
1.	Ri- Bhoi	26	156	6000
2.	East Khasi Hills	11322	120013	10600
3.	West Khasi Hills	5849	41100	7027
4.	Jaintia	284	2088	7352
5.	East Garo Hills	118	917	7771
6.	West Garo Hills	423	3228	7631
7.	South Garo Hills	49	382	7796
	Total Meghalaya	**18071**	**167884**	**9290**

Traditional Potato Growing Practices Followed in Meghalaya

Land Preparation Method

The state of Meghalaya consists primarily of mountainous

terrain and different crops including potato are cultivated predominantly on the slopes. Land preparation for potato cultivation in the Meghalaya state is carried out in indigenous or traditional way which is locally called as '*Bun and Nur* method' wherein the '*Nur*' means the local land unit *i.e.*, one raised bed and '*Bun*' is the method of preparation of these raised beds. This method is an important form of shifting or *Jhum* cultivation which is in practice among the tribal farming community in the state since time immemorial. *Bun* and *Nur* method is basically a type of ridge and furrow method which has been modified to suit the difficult mountainous hilly terrain existing in the state.

This method of land preparation *i.e.*, *Nur* preparation, is followed whenever a new forest or fallow area, etc is to be brought under potato cultivation. In this method of land preparation for potato cultivation, rows of raised beds each measuring around 1-1.25 m wide, 20-25 cm in height and 2-7 m long (depending on the availability of land), are made along the slope. The distance between two beds is kept about 50-70 cm, which works as drainage channel to drain excess of rain water.

These raised beds '*nurs*' are prepared differently for summer and autumn potato planting or in other words different forms of bun system are practiced for land preparation in summer and autumn season. Though these methods are not mutually exclusive, however need to be mentioned in separate heads. These are presented in the following paragraphs :

(*i*) *Land Preparation for Summer Potato Planting*

(*a*) Land preparation starts by cleaning the area from the grass, shrubs, fallen pine leaves, twigs, branches, etc. These along with sods are then spread on the raised beds. Some of the farmers also practice making heaps of these in the drainage channels in the field. The collected matter is are then turned over with spade and exposed to sun for sometime to get dry and during the month of February and March (at planting time) it is burnt. In case where twigs or tree branches are used, they are covered with 10-15 cm thick layer of soil before it is burnt. Soon after the soil cools down from burning, the planting is done. This method of land preparation is the bun method. '*Bun*' the word literally means 'materials for burning' in the local khasi language. The land prepared by following the procedure is utilized for cultivation for 2-3 years or till it is realized by the farmers that soil fertility has gone down, and then the land is left fallow for the next 2-3 years or more. The grass and shrubs that grow on this fallow land during this period is utilised for making bun or burning for preparing land for potato cultivation.

(*b*) Another method of land preparation followed in the state is more closer to the widely known form of Jhum or shifting cultivation than the above mentioned bun method. In this method during February and March (potato planting time) the trees and shrubs are chopped off from the entire forest area selected for potato cultivation and is burnt. Once the fire extinguishes, raised beds similar in measurement as mentioned in above method are made along the slope with the drainage channels. A cycle of five years is followed for cultivation in this method of land preparation.

(ii) Land Preparation for Autumn Potato Planting

For autumn planting, modified form of above mentioned methods of bun preparation are utilized. In this method FYM is not utilized for planting. These are mentioned hereunder :

(*a*) The first method of nur preparation includes cutting of long grasses from the bed/land and nearby areas and then clearing the land/beds from the left over grass/ stubbles along with the soil (thus get sods). The grass and sods are kept in upright position (soil facing downwards) on the cleaned beds followed by keeping seed tubers and then covering all with soil. Use of FYM is not practiced in this method of land preparation.

(*b*) In this second method of land preparation for autumn planting, the cut grass and sods are kept first on the cleaned bed/ land to be utilized for potato planting. A mixture of fertilizers (Urea + Single Super Phosphate) is then sprinkled on the grass and sods laid on beds and finally beds are covered with soil and kept as such for a period varying from of 2-3 days to 3 weeks depending upon the availability of time with the farmer for actual planting, the congenial weather conditions, harvesting of earlier crops, availability of seed, labour, etc. Farmers however, prefer a longer period for this method of planting. After this period, shallow pits are dug on the beds and potato is planted therein.

Nurs or land prepared following this way is utilized for the autumn crop and also for the summer season potato crop continuously and then after the summer season produce is harvested, the land is kept fallow for 2-3 years and then again is prepared using one of the above mentioned methods for taking any crop. However, potato growers with limited land continue to use the same nur/ land for further seasons.

Farmers' Rationale of Traditional Methods of Land Preparation

(*i*) ***Cultivation on Raised Beds made Along the Slope:*** Farmers perceive that cultivation of potato and other vegetables on raised bed help them in preventing the washing off of the entire crop during heavy rain and storms. The drainage channels laid along with these beds help in draining the excess rainwater without affecting much the crop on the raised bed.

(*ii*) ***Bun Method of Cultivation :*** Potato growers feel that burning involved in bun method of land preparation helps them to cultivate the hard and compact forest soil easily. The 20-25 cm deep soil which is the height of the raised bed, is made available for cultivation, which otherwise would be labor some to get if forest earth is to be dug and used for planting. The soil in this method becomes light, friable porous with high water holding capacity as the ash of the materials burnt i.e., twigs, shrubs, leaves, grasses etc., is added to the soil. In addition, the incidence of weed and also disease is perceived to be low in this method of land preparation thus leading to a better production.

(*iii*) ***Keeping the Land Fallow for 2-3 Years :*** Farmers believe that if the land is cultivated continuously for some years, land/sod becomes very soft and the soil fertility is reduced, thus rendering the land unfit for cultivation. Leaving land fallow allows it to get harden for cultivation and also rejuvenate naturally.

(*iv*) ***No Use of FYM in Autumn Planting*:** Farmer perceive that if FYM is used in autumn planting, the heavy rain in the month of July to October would make FYM wet and which in turn would lead to rotting of seed tubers in field, especially if the rain continues for longer duration. In addition, farmers perceive that the grass and sods used in land preparation would get decomposed and act as manure for the crop.

Land Ownership Pattern

Land ownership pattern in the state of Meghalaya is quite unique in India. Majority of the agricultural land in the state is owned either by certain clans or it is community owned or in few cases owned by individual cultivators. The landless cultivators then leased in the land from the clans on yearly basis. In case of community owned land, the village headman allocates the land to the cultivators and in turn collects a minimal rent from them.

Seed Preparation/Management

(i) Seed Source

Farmers of Meghalaya utilize both the formal and informal seed source, which include the state department of horticulture, private seed suppliers, fellow farmers and farmer's own seed. Using their own seed for the subsequent crop and borrowing seed from fellow farmers (whose crop farmers perceive to be healthy) are the preferred practices among majority of the farmers in the state. Formal Network of Seed distribution by Meghalaya State Agricultural Department is though functional but is limited in coverage in terms of area as well as numbers of farmers.

Farmers' Rationale/Reasons

All potato growers in the state are not aware of alternate seed source (the formal seed source) operating in the state. In addition, farmers perceived that state department does not have adequate seed to supply to all farmers. Using the own seed is also preferred as compared to the supply by private seed trader by the potato growers in the state due to the reason that the seed tubers supplied by later them usually are not uniform in size, contain varietal mixture and include damaged tubers also.

(ii) Seed Use Pattern

Whole potato tubers are planted as seed by majority of the farmers in Meghalaya. While some of the farmers practice use of cut tubers. Few farmers also utilize a mixture of both the method *i.e.* cut as well as whole tubers. It was found that farmers practiced the use of cut tubers only in case when the tuber size is large. Such large sized tubers are usually cut vertically into two pieces. None of the surveyed farmers was observed doing the practice of treatment of seed (either cut or whole) before planting. Farmers in the state utilize all sizes of healthy seed tubers for planting in the state, irrespective of the size of the tuber. Seed replacement in the state is usually not practiced by majority of the farmers. Few of them who follow the seed replacement do it once in every 4-6 years.

Farmers' Rationale/Reasons

Lack of awareness of the benefits of treatment of cut tubers, the benefits or the methods of treatment of seed tubers before planting among the potato growing farmers in the state. Farmers continue to use their own produce as seed for potato cultivation as they lack awareness about the benefits of replacing the seed and also quality seed for replacement is not made available in adequate quantity by the state agricultural department.

(iii) Seed Rate and Planting Distance

Potato growing tribal farmers in Meghalaya state prefer to use higher seed rate ranging from 35 q/ha to as high as 60 q/ha. Accordingly the inter-row and inter-plant distance is also kept to a minimum. The spacing also varies from season to season. In summer season, the row to row spacing is kept around 20-30 cm and plant to plant is kept about 20 cm, while for the autumn season it is on still lower side, *i.e.* 15-20 cm row to row and 15-20 cm plant to plant.

Farmers' Rationale/Reasons

Farmers' idea behind the use of minimum spacing and in turn higher seed rate in the state is the maximum utilization of the available land. The land is a limiting production factor in this hilly state. In addition, the land ownership pattern followed in the state wherein the cultivators lease in/rent in land from the landowners, also promote such practice of using a higher seed rate and a lower spacing with a perception that a higher production would be possible.

Planting Method

Planting of potato is practiced in pits dug on the raised beds '*nurs*'. For summer planting, once the pits are dug on the nurs, the seed tubers (one each) are kept in these pits that is followed by putting FYM over the seed and then fertilizer mixture on the outer edges of the pits and finally the pits are covered with soil. For autumn season a variation is observed in the above-mentioned method of planting. Some farmers practiced use of the summer method of planting while other used a different method, which is mentioned in 3.1 subhead.

Rationale / Reasons

The practice of pit method of potato planting prevents drying of seed tuber especially in case of early potato planting even if there is a dry spell for sometime after the planting. In addition, farmers perceive that the said practice also provides seed tuber protection against insect and pest. Moreover, placement of fertilizer mixture over and around the FYM help the later in early/easy decomposition and converting it into the available form for the tubers.

Use of Fertilizer and FYM

Use of urea and Single Super Phosphate (SSP) are in

practice in the state, however, farmers did not apply Potash for their crops. Farmers do liberal use of FYM for the potato crop. A variety of manure is utilized for the crop including cow dung, pig faeces, poultry faeces and decomposed grass. Farmers apply manure only for the summer potato crop in the state. The rate of manure is also varied according to the land type. For upland, the use of FYM vary from 250-350 q/ha and for low land, manure is applied at a higher rate *i.e.* 500 q/ha.

Farmers' Rationale/Reasons

Non-availability of potash is the reason cited by farmers for not applying it to the potato crop. Also, the manure applied at the time of planting is usually wet and this in turns increases the weight of the manure. In addition, farmers perceive that use of fertilizer, which is locally called "*Dawai*", is harmful for the land fertility in long run and also for the human health, therefore they prefer use of higher dose of FYM.

Interculturing Operations

Potato growers practiced weeding and earthing up of potato crop when crop attains a height of 5-7 inch in both the potato growing seasons. Top dressing with Urea @ 200 gm/nur, however, is practiced only for autumn crop. Top dressing is not practiced for summer potato crop.

Farmers' Rationale/Reasons

Farmers practice top dressing with urea in autumn crop because the area planted for autumn crop is less as compared to the area in main summer season therefore buying urea for smaller area involves spending less money. Also, farmers feel that heavy dose of FYM in summer crop continue to supply nutrients to the crop throughout its growth stage.

Management of Major Diseases

Late blight and bacterial wilt/ brown rot were identified by tribal farmers as the major diseases affecting potato crop in the state. A description of identification of symptoms and management practices followed by farmer related to these diseases are mentioned in the following paragraphs.

(i) *Late Blight*

Identification of Symptoms : Late blight is regular and endemic in nature in the state locally called as 'Yapyong'. This disease is identified by symptoms like turning of leaves from green to brown or black in colour, drying of leaves and development of white cottony growth on underside of leaves.

Extent of Damage : Farmers perceived that late blight though causes damage, however, in both the potato growing season, i.e. summer and autumn, the extent of damage is higher in autumn potato crop. It is also perceived that incidence of late blight is higher in upland areas (slopes) as compared to the low-lying areas in valleys.

Management of Late Blight : Majority of the potato growing farmers in the state were observed to practice spray of fungicide (Indofil) @ 5 teaspoons in 15 liters of water. However, the spray is restricted only to the autumn crop (July-Aug-Sep). Only few farmers practiced spraying of fungicide in summer crop also. Some farmers have also devised the cultural practice of taking early potato crop wherein planting is done in February and harvesting in the month of June giving almost four months for the crop to mature. This way the effect of blight which initially appears in last fortnight of May is minimized/escaped.

Farmers' Rationale/Reasons : Farmers did not practice spray of fungicide in main summer crop season owing to the preference among them for harvesting potato crop early so as to prepare land for the second crop, which is either cabbage or cauliflower in slopes (upland) and paddy in low land. Farmers perceived that spraying of crop with fungicide would lead to delay in maturity of the crop and thus its harvesting will not fit and would be unfavourable for the cropping sequence followed by them.

(ii) *Brown Rot*

Identification of Symptoms : Brown rot was found identified by the potato farmers as a disease that occurs in store only leading to rotting of seed tubers. In addition, appearance of brown ring in cut tubers is also taken as symptom of brown rot. However in field the extent of occurrence was reported less.

Extent of Damage : Farmers perceive that the extent of damage due to brown rot is more in produce of autumn crop and its incidence is higher in crop harvested from low lying areas in comparison to that from upland or slopes.

Control/ Management Measures : Manual sorting of diseased tubers at the harvesting and storing time is the only management measure for brown rot followed by farmers of the state. Interestingly, farmers did not perform sorting of diseased tubers in storage. Keeping the stored seed tubers undisturbed (not performing sorting) is another practice followed by farmers to check the spread of this disease..

Farmers' Rationale/ Reasons : Farmers perceive that touching or sorting the stored seed tubers causes injury to the tubers and enhances the disease spread.

(iii) Bacterial Wilt

Identification of Symptoms : Bacterial wilt was identified by majority of the farmers by observing the drying of stem and leaves followed by drooping of entire plant. Some of the farmers also identified drooping leaves as the early symptom of bacterial wilt in crop.

Extent of Damage : Higher incidence of bacterial wilt was perceived by farmers to occur in autumn crop in comparison to summer crop and also in low-lying areas as compared to the crop in slopes.

Control / Management Measures : Rouging of diseased plant is practiced by some farmers in state for management of the bacterial wilt disease.

Insect Pest Management : Farmers revealed that white grub, cutworms and Potato Tuber Moth (PTM) are the major insect pest damaging the potato crop. A description of the identification symptoms, extent of damage and control measure adopted by the farmers are given in the following sub-heads.

Identification of Symptoms : Infestation of white grub is identified by observing damage of tubers in fields. Cutting of stem in standing crop and observation of holes in tubers was reported the by farmers as the infestation by cutworms, white grub respectively. Damage by Potato Tuber Moth was identified with damage to potato eyes in stores.

Extent of Damage : Farmers perceived that these pest caused more damage to summer crop as compared to autumn crop may be because the fact that during the maturing and harvesting of potato in summer crop the temperature is relatively higher as compared to that prevailing during maturing and harvesting of autumn crop.

Management Practice : Use of chemical/insecticide for management of these insect pests are not in practice by most of the farmers in the state. However, in case of Potato Tuber Moth (PTM) management, few farmers used sometimes insecticide in the form of dust in the store. Manual sorting of damaged tuber is found frequently followed for management of cutworm as well as white grub. Some farmers in the state are also observed using chopped dried *Lantana* leaves by spreading over the tubers in store for management of PTM.

Farmers' Rationale/Reasons : Farmers of the state perceived that the '*Bun*' method of land preparation helped them to prevent heavy infestation of insect pest on crop. They also perceived that additional economic burden would be involved in potato cultivation if high quantity of insecticides will be used against the relatively lesser extent of damage by insect-pest and hence, their profit margin in potato crop may be reduced which is also becoming less remunerative in recent years.

Harvesting

Harvesting time of potato crop varies from the month of June to August depending on number of factors. These factors include the time of planting, the next crop to be grown, market price of produce, etc. Harvesting is usually preferred in bright sunny days. It is primarily carried out by women folk with men helping in carrying produce from the fields to the stores

Harvesting is done with the locally made small spade called "*Moh Khiew*" in Khasi. The harvested produce is carried in bamboo baskets with the help of sling. The bamboo baskets are carried by men folk on their back with the help of sling, the loop of which is made to hang from the head to get support.

Farmers' Rationale/ Reasons : The reason mentioned by farmers' for harvesting of potato crop over a long period of time is the market price. During the months of June – July when the crop matures in the state the markets are flooded with the potatoes imported from the neighboring states at a cheaper rate and also the fresh harvested produce from within the state is also arriving in the market leading to a lowering of the potato price in market. Therefore, during these months potato planted in the low land is harvested as the land is to be prepared for paddy crop. For the upland areas, some farmers let the crop to store in the field itself and harvesting is not done at all or in other cases the crop is harvested and the produce is stored for a few months untill the flow of potatoes in the market from within the state decreases. The excess of produce which their stores can not accommodate is sold in the market immediately after harvest.

Grading

Potato produce harvested by the farmers that is meant for sale in the market is generally graded into large and small size only. Potato that is to be used for seed as well as table purpose for home consumption is generally not graded. For seed purpose the healthy and disease free tubers are selected, irrespective of the size. However, too small and too large size is usually avoided by the farmers for seed purpose.

Potato Storage

Unique indigenous potato storage structures as well as methods are being utilized by the farmers in the state. These storage structures/methods can be classified broadly into two types- the *In-Situ* and *Ex-Situ* storage. A description of storage structures and methods and the different types within these are mentioned in following subheads.

(i) In-Situ Storage (in ground storage/delayed harvesting)

This is the type of storage system in which the produce

is stored in fields itself and not harvested even after the crop matures and dies. However, such type of storage is utilized for short term storage i.e., for 3-4 months only. Two types of In-situ storage are practiced by farmers in the state.

(*i*) In the first type of storage which is usually practiced in case where potato crop is taken on hill slopes, is that the harvesting of potato crop is delayed. Consequently, due to heavy rainfall in the following months, grass grows on these potato beds. The grass provides cover or shade to the potato beds and the potato stored therein. This allows the temperature inside the soil to be a bit lower as compared to the warm temperature existing outside.

(*ii*) In the second type of storage, which is usually practiced by the potato growers in the state when potato is planted in low-lying areas, is that the paddy crop is taken on top of beds wherein potato is kept for storage. In this case, the paddy crop provides shade and cooling affect to the stored potato inside the soil. However, this type storage is presently being practiced by a limited number of farmers in the state.

Farmers' Rationale / Reasons : Farmers perceived that the rottage of potato tubers in such *in-situ* storage method is less in comparison to potato stored in country made storage structures. In case where paddy crop is used for cover, farmers perceived that the paddy crop is healthier.

The time of potato harvesting was also determined by the market price. During the month of June- July i.e. usual harvesting time for potato in the state, low market price prevail, therefore farmers prefer to store a part of their produce for some months and wait for market price to rise.However, during the same period (during June-July) the climatic factors like the heavy rainfall as well as high temperature does not permit farmers to store all the harvested produce for long period in their own indigenous stores as rottage is high in such climatic conditions. In addition, space for storing all the harvest is also a constraint, which favour *in-situ* storage of the produce.

(ii) Ex –Situ Storage Method

Various types of indigenous potato storage structures are

being utilized by farmers in the Meghalaya.A brief description of these is made in following sub-heads:-

Pit Method of Storage : In this method of storage, a underground circular pit of is dug and seed potato are kept in it. The tubers are then covered with soil and finally with straw on top. These pits are dug with varying dimensions. These storage pits are dug on an elevated area or ground. This method of storage is used for storage of seed potatoes only.

Pit Storage in Layers : In this method, a large pit is dug and layers of 1-1.5 inch thick layers each of straw, seed, straw and soil are made one on top of another. These layers are repeated for 4-5 times. The pit is then provided shade with a structure made of bamboo which is covered with thatch on top.

Wooden Storage Structure : Potato growers in Meghalaya store the potato harvest in the wooden storage structures that are built in the field or near residential area. These structures are made of wooden planks nailed into each other. The planks are nailed horizontally on the four support structures (forming the four corners of store) so that the second planks overlaps the first one and simultaneously the 3rd overlapping the second one and it continues till the desired height of 10 feet is reached. The structure thus made is covered on top with tin roof. A considerable gap is left between the roof and walls thus made, giving enough space for aeration in the store. These types of wooden structures are one of the most popularly used potato storage structures among the farmers in the Meghalaya.

Storage Rooms : During the potato harvesting time, farmers turn a part of their house as store. Usually a tiny room on the ground floor is converted for storing the produce. These converted rooms are usually without proper light and aeration. This type of storage is utilized for seed as well short term storage of produce that is to be sold in market, once the market price rises.

Storage Space below Houses : Houses in Meghalaya are constructed so that they are slightly raised from the ground supported by four or more pillars from below. The space below the houses are also used by farmers temporarily to store the potatoes. The floor of this store space is earthen and also damp which leads to heavy rottage.

***Pucca* Storage :** This type of storage structures are either made of cement and stone or of Cement and bricks. However, fewer farmers have such types of storage structure owing to the cost factor.

Farmers' Rationale / Reason : Local availability of the materials for making these structures, low cost or in some cases no additional cost in constructing these storage, the local skill available for constructing these structures were the reasons cited by farmers for preferring these potato storage structures. Limitation in terms of available space was another important reason cited by farmers for utilizing any available space for the purpose.

(iii) Storage Systems

Storage in Baskets : Bamboo baskets locally called as "*Polo*" is usually used by farmers in the state to store potato. Different sizes of polo are utilized by farmers ranging from basket that can hold 10-12 kg potatoes to one that can hold approximately one-quintal potato. Farmers prefer smaller baskets for storing seed potato. Farmers use this method of storage for both short and long term storage i.e., for table and seed purpose. However, in certain areas of the state, seed potato is preferred to be used in these baskets. This is one of the most popular methods of storage that is practiced by a large percentage of potato growers in the state.

Farmers' Rationale / Reasons : (*a*) Farmers perceive that storing potatoes in bamboo baskets allows proper aeration, thereby cause less rottage. In addition use of bamboo baskets is very popular among the farmers for storage of potato owing to the reason that bamboo is a locally available and therefore is cheap.

(*b*) Storing of seed potato is preferred by farmers in small baskets with the rationale that when less layers of potato are stored together tuber rottage is very less and also small baskets are easy in carrying them to the fields at planting time.

Storage in Gunny Bags : Some farmers in the state also practice storing the potatoes in gunny bags, which are usually kept inside the stores.

Storage in Heaps

Potato storage in heaps inside the storage structures is also in practice by farmers. This form of storage is practiced for seed as well as table potatoes.

Farmers' Rationale / Reason : Limitation of storage space is often cited as reason for following this storage by the farmers of the state.

Storage in Layers in Different Platforms

In this type of storage, different platforms are constructed in the stores that are either made of bamboo or wooden planks. In this form the potatoes are spread on these platforms in thin layers. This form of storage is used for both tables as well as seed potato. This form however, is a recent introduction in the state that is becoming popular in certain pockets in the state.

Farmers' Rationale / Reason : The rationale cited by farmers for following this method of storage is that this type of storage provides a better space utilization and also that the tuber rottage is less.

Breaking Seed Dormancy

Potato growing tribal farmers in Meghalaya has evolved a unique way to break the seed dormancy. This method is being utilized by farmers for using the produce of summer harvest and also for late harvested autumn potatoes as the seed for the subsequent autumn and summer season crop planting for which starts in Mid July to August and early February, respectively. In this method the seed tubers are spread in thin layers on dry surface (floor). This is then covered with dry grass or old gunny bags to facilitate sprout growth. This method of breaking seed dormancy is presently being frequently used by farmers for the summer season harvested crop of Kufri Giriraj, which is again utilized as the seed for the autumn season crop.

Scientific Evidences Supporting Farmers' Rationale of Traditional Potato Cultivation Practices

- Farmers believe that the *Bun* method (i.e. soil burning) helps in improving soil fertility status and makes hard forest soil friable, thereby, making potato cultivation easy. This notion is supported by findings of Griffith (1947) and Gouroup (1957) who reported improvement in soil physical condition on burning. An increase in soil fertility status is also reported by several researchers. Nyle and Greenland (1960) reported that burning of forest considerably increased phosphate and exchangeable cation content of the soil. The potato yields were also significantly increased. The increase was comparable to that obtained with application of 100 kg N, 65.5 kg P_2O_5 and 41.1 kg K_2O per hectare (Awasthi *et al.,* 1981). Thus by soil burning the use of fertilizers for the potato crop could be avoided in the first year.
- The belief that incidence of diseases in the crop is less in *bun* method of reported that, besides direct effect of soil burning, the smoke liberated in the process killed a number of plant pathogens. Similar results were reported by Hardison (1976). Awasthi *et al.* (1981) reported that soil burning in *bun* method completely eliminated the soil inoculum of brown rot (*Psedomonas solanacearum* Smith) in the first year. Soil burning increased the soil pH from 5.1 to 5.5 (Anon, 1981). An increase in the soil pH was also reported by Mukherjee way back in (1954) and Nyle and Greenland (1960).
- The reduced incidence of insect pests in *bun* method of potato cultivation may also be due to cutting and burning of plants. This destroys well-laid habitat, food supply and specific environment of insect pests. Insects from nearby areas are also attracted by the light (fire) and killed.
- The *nur* and *bun* method of planting used by farmers to protect the crop from washing away in heavy rains, however, causes heavy soil erosion in the hill slopes. It is estimated that for every single ton of potato harvested from a field on 38% hill slope, two tons of soil is lost (Singh and Singh, 1981; Mishra and Gupta, 2001).

Integration of Potato in the Productions, Consumption and Disposal Pattern

This subhead presents the reasons supporting the prevalence of potato based cropping system, varietal preferences of the farmers, consumption and disposal pattern of potato followed in the state of Meghalaya.

Potato based Crop Sequences :

Potato forms an integral part of major cropping sequence followed in the state specially the major potato-growing region in the state. The major crop sequences practiced are-

(*i*) Potato (early summer)-Paddy-vegetables.

(*ii*) Potato-cabbage/Cauliflower-Radish/Beat

(*iii*) Bean-Potato (Autumn)

(*iv*) Potato (Summer)-Potato (Autumn)

(*v*) Potato (early summer) - Maize

Farmers' Rationale / Reason : Short duration of the potato crop that gives enough time for taking another vegetable crops, better storability of potato as compared to the other vegetables, a good production of potato and simple cultivation practices are some of the reasons mentioned by potato growing farmers for preferring potato crop in the state. Moreover, the climatic conditions favour potato cultivation in the two consecutive seasons.

Potato based Intercropping

The most popular potato based intercropping followed with potato crop in the state is with the maize. In some places in the state however, potato crop is also intercropped with Pea or *Sohphlang* or Soybean or Yam.

Varieties Cultivated and Relative Preferences

Locally available varieties like *Lah Shaw, Lah Syntiew, Lah Lupon* etc are still in cultivation in the state, though these are cultivated on a small scale particularly for home consumption. The areas under these local varieties are slowly being replaced by the high yielding cultivars like K.Jyoti,

K.Giriraj, and Kufri Megha. Majority of the potato growing area in the Meghalaya state is under Kufri Jyoti. Kufri Megha is a variety, which is found still popular among farmers in the state. Kufri Giriraj is a recently introduced variety in the state and is becoming popular in the state.

Farmers' Rationale / Reason : The local varieties though usually poor yielder in comparison to the other high yielding cultivars, however they are still in cultivation in the state due to the taste factors and better cooking quality. Farmers in the state perceive that though K.Megha is slow bulker, but due to its high resistance against late blight disease, it gives low but assured yield to the farmers.

Kufri Jyoti variety is slowly loosing its resistance against the late blight disease in the state leading to a reduction in production and yield. Therefore, potato growers in some pockets prefer the newly introduced cultivar-K. Giriraj which is becoming popular in the state owing to the basic reason of high resistance against late blight as well as higher yields. Farmers in the state however are still skeptical about its performance owing to the poor keeping quality of this variety and the short dormancy.

In addition, the other reason for the K. Giriraj becoming popular in the state is that the shortcoming of the K.Giriraj variety i.e. short dormancy is being exploited by the farmers to their benefit because of prevalence of two potato growing seasons in the state i.e. summer and autumn by way of using the produce of summer crop of as seed for the autumn crop planting.

Potato Consumption Pattern

Potato is a preferred as vegetable among the people of Meghalaya state. This vegetable is consumed as a part of all the three meals of the day. People's preference for consuming potato is by mixing with other vegetables, consuming it as sole vegetable or after boiling. Some people also consume potato after roasting it. It could also be noted that the state of Meghalaya has highest per capita potato consumption (65kgs/ person/ year) compared to the national average of 17 kgs/

person/ year. Potato tubers of bigger size are more preferred by people of Meghalaya state for consumption.

Farmers' Rationale / Reason : Consuming large sized potato is preferred by people as the fried potatoes are the most common preparation and also people perceive cooking of larger tubers easier.

Potato Disposal Pattern

Majority of the farmers in the state prefers selling large sized potato tubers direct to the retailers in the city vegetable market. They also dispose their produce to the fellow farmers, if they get a request for it. The small sized tubers as are not preferred by people in the state for consumption and hence donot fetch a good price in the local market. Owing to this fact, these are either sold to the middleman who comes to collect the produce from the villages at a minimum price or farmers directly sell the small sized tuber to the potato traders in Shillong market or farmers sell the small sized tuber in the regulated market. Some of the potato growing farmers in the state also practice selling a part of their produce in the local village markets.

Farmers' Rationale / Reasons : Preference for bigger sized tubers for home consumption among the people of Meghalaya lead to higher price for the big sized tubers in the local vegetable market, thus farmers in turn also get a better price for their produce if they sell their produce to the retailers. Traders on the other hand, purchase the small sized tubers from the farmers for the purpose of selling it in the neighboring states wherein the small tuber size is preferred for table purpose and also as seed.

Analysis of Technological Gap in the Adoption of Improved Potato Production, Protection and Storage Practices and Intervention Required

As deliberated in the preceding subheads, it was found that majority of the potato farmers in the study area were resorting to the traditional methods of potato production, protection and post-harvest handling of potato. It was therefore pertinent to analyze the extent of the technological gap in the adoption of scientific method of potato cultivation and at the same time to ascertain the nature of interventions required to surmount the existing gap.

Table 2.3 : Technological Gap in the Adoption of Recommended Potato Production Technologies (n = 120)

Sl No	*Improved Practices*	*Techno logical Gap(%)*	*Interven tion required (Yes/No)*	*Nature of intervention suggested*
1.	Replacement of seed every 3-4 year	100%	Yes	On-farm assessment of quality seed
2.	Chemical treatment of seed before planting	100%	Yes	On-farm trials
3.	Seed rate	100%	Yes	On-farm demonstration
4.	FYM application	—	No	–
5.	Planting method	100%	Yes	On-farm trial on Recommended methods vs traditional method.
6.	Balanced use of chemical fertilizer	80%	Yes	On-farm demonstration on balanced use of chemical fertilizer
7.	Management of late blight	50%	Yes	On-farm trial on use of Ridomil along with Dithane M-45
8.	Management of bacterial wilt	100%	Yes	On-farm trial on chemical control and use of disease free seed.
9.	Management of insect pest	100%	Yes	On-farm trial on use of insecticide.
10.	Sorting and grading of tubers.	–	No	–
11.	Storage of harvested produce	80%	Yes	Trial on low cost storage structure

The findings as presented in the enclosed Table 2.3, it could be seen that there was 100% technological gap in the adoption of major practices like replacement of seed every 3-4 year, chemical treatment of seed before planting, seed rate, and planting method. In addition, the technology gap of the same magnitude was also found in case of adoption of management of bacterial wilt and management of insect pest. However, with respect to the practices like balanced use of chemical fertilizer, management of late blight and storage of harvested produce the gap was computed to the relatively lower extent varying from 50%-80%. Similarly, the gap in the adoption of improved potato cultivar was found to the lowest extent of 25%. Similarly, in case of practices like FYM application and sorting and grading of harvested tubers, gap in the adoption was noted as zero.

From the above findings, it could be observed that that recommended technologies that were relatively difficult and complex to perform and at the same time were not compatible to their existing biophysical situations were not adopted by the tribal farmers. Under this case, therefore, it was found that either on-farm trials could be the appropriate persuasive tools for such category of farmers. Similarly, as there was application of high dose of FYM in potato prevailing among the farmers and also there was use of HYV of potatoes mainly K. Jyoti and K. Megha as well as K. Giriraj (recent introduction), farmers were in use of recommended application. Similarly, for management of Late Blight, farmers were adopting the spray schedule especially for autumn crop. However, in case of these technologies also, it was found that on-farm demonstration could be the suitable tools for further sharpening the potato farmers regarding accurate application of these technologies.

Need Perception of Tribal Farmers With Regard to Recommended Potato Production Practices in Meghalaya

"Need" refers to the positive driving force that impels a person towards certain objectives or conditions. Needs are operationalised as the training and infrastructure needs as perceived by the potato growers with regard to recommended scientific potato production practices. Training needs of tribal potato growers therefore are identified in the broad areas of the recommended potato production practices i.e. potato production, plant protection, post harvest handling etc. In addition, infrastructure needs are also ascertained on potato production practices in the state. Needs are assessed for individual training and infrastructure need item on a three-point continuum i.e. urgently needed, needed and not needed and accordingly the score of 2,1 and 0 respectively was assigned. Need index for Individual training need was calculated by using the following formula.

$$\textbf{Need Index} = \frac{\text{Total obtained score of the total respondents}}{\text{Total obtainable score for all the respondents}} \times 100$$

Training needs related to recommend Potato Production Practices

Training Needs Related to Potato Production : From the Table 2.4, it could be observed that among all the training need items relating to potato agronomy, scientific method of potato planting is perceived as highest followed by balanced use of chemical fertilizer and land preparation respectively. The tribal farmers, though, perceived training need on items like soil testing for fertility status and application of FYM but with lower intensity.

Planting of potato is carried out by the tribal farmers of the state in the traditional pit method of raised bed that is highly labour intensive with use of a higher seed rate. In addition, the intercultural operations like weeding and earthing up are performed by the farmers in the traditional manner only. These could be probable reason for majority of tribal farmers' of the state perceiving the training need on scientific planting method. Another characteristic feature of potato cultivation in Meghalaya is the heavy use of FYM along with minimum use of chemical fertilizer, which also explains the high training need index with respect to the item balanced uses of chemical fertilizer along with a low need with respect to application of FYM.

Table 2.4 : Training Needs Related to Recommended Potato Production Practices (n = 120)

Sl. No.	*Training need area*	*Training need index*						*Average*	*Over all rank*
		C1 (n1=20)	*C2 (n2=20)*	*C3 (n3=20)*	*C4 (n4=20)*	*C5 (n5=20)*	*C6 (n6=20)*		
1.	Land	37.50	45.00	62.50	47.50	25.00	32.50	**41.67**	**III**
2.	Preparation	37.50	72.50	60.00	42.50	72.50	55.00	**56.67**	**I**
3.	Scientific Method of Planting	50.00	15.00	12.50	17.50	15.00	20.00	**21.67**	**IV**
4.	Soil Testing for fertility	10.00	27.50	35.00	20.00	20.0	20.00	**22.92**	**V**
5.	Application of FYM Balanced use of chemical fertilizer	70.00	57.50	47.50	50.00	47.50	47.50	**53.75**	**II**

C1 = Circle 1, C6 = Circle 6 (as there are six potatao growing circles in Mcghalaya)

Training Needs Related to Plant Protection Measures : Among the four items pertaining to plant protection, training need is perceived highest in management and control of late blight with need index of 72.08, followed by management of brown rot for which need index was 62.92 by the tribal farmers of Meghalaya. While for item like chemical treatment of seed before planting the seed index are only 44.17. The training need was found lowest on the item of insect pest identification and control. (Table 2.5)

Late blight is a regular occurring disease in the potato crop in the state that affects both the summer and autumn crop in devastating manner. This could be the reason for the item management and control of late blight acquiring highest priority by respondents for training. Brown rot is another serious problem affecting potato in the field as well as in store. Insect pest identification in the state is less apparent except during dry warm season in comparison to the devastation by diseases. This could be the probable reason for lesser respondents perceiving the training need on insect–pest identification and control.

Table 2.5 : Training Needs Related to Plant Protection Measures

Sl. No.	*Training need area*	*Training need index*						*Average*	*Over all rank*
		C1	*C2*	*C3*	*C4*	*C5*	*C6*		
1.	Chemical treatment of seed before planting	45.00	60.00	25.00	27.00	52.50	52.50	**44.17**	**III**
2.	Management and control of late blight.	67.50	82.50	67.50	62.50	80.00	72.50	**72.08**	**I**
3.	Management and control of brown rot	77.50	75.00	37.50	70.00	60.00	57.50	**62.92**	**II**
4.	Insect pest identification and control	40.00	50.00	55.00	35.00	30.00	40.00	**41.67**	**IV**

C1 = Circle 1, C6 = Circle 6

Training Needs Related to Post Harvest Handling: Grading and sorting of potatoes and storage of potatoes are the only two identified training need items in this post harvest handling aspects. On closer look at the training need index obtain for this area, it could be seen that the index are lower as compared to the need index of items in other subheads. Among the two post harvest handling items training need of respondents was found to be higher for item on proper storage of potato (46.25) and for the other item of grading and sorting of potatoes the training need index was found to be low i.e. 15.42 only. (Table 2.6)

Table 2.6 : Training Needs Related to Post Harvest Handling of Potatoes

Sl. No.	*Training need area*	*Training need index*						*Average*	*Final ranking*
		C1	*C2*	*C3*	*C4*	*C5*	*C6*		
1.	Grading and sorting of potato	30.00	17.50	15.00	10.00	12.50	7.50	**15.42**	**II**
2.	Proper storage of potato	65.00	55.00	32.50	32.50	45.00	47.50	**46.25**	**I**

C1 = Circle 1, C6 = Circle 6

Potatoes in the state of Meghalaya are mainly stored in traditional country made stores which are though not scientific but are constructed on the available space, with the available skill, material and methods. The tribal farmers in the state practice grading and sorting of potatoes based on their own perception of size of tubers. These facts could be the probable reasons for the items of Proper storage of Potato, receiving a high training need index and grading and sorting item receiving lesser priority.

Infrastructure Needs

Availability of adequate infrastructure facilities like shops for supply of fertilizer, transportation, motor able roads, etc is one of the important factors that may determine the acceptance of recommended/improved potato production practices among the tribal farmers of the state who are operating in a hilly and difficult terrain. An attempt has therefore been made to analyze the infrastructure requirement in the study locale as perceived by the potato growers.

Table 7 : Infrastructure needs related to Potato production, storage and marketing.

Sl. No.	*Training need area*	*Training need index*						*Average*	*Final ranking*
		C1	*C2*	*C3*	*C4*	*C5*	*C6*		
1.	Provision for well-knit seed supply agency	62.50	70.00	40.00	60.00	40.00	52.50	**54.17**	**IV**
2.	Cooperative store for availability of fertilizer at appropriate distance	85.00	62.50	50.00	47.50	82.50	70.00	**66.25**	**I**
3.	Govt. shop for supply of fungicides, insecticides etc at appropriate distance	72.50	52.50	60.00	70.00	35.00	62.50	**58.75**	**III**
4.	Facility of cold storage	37.50	42.50	67.50	40.00	57.50	32.50	**46.25**	**V**
5.	Facility of country made storage (ware house /godowns)	45.00	25.00	27.50	25.00	75.00	42.50	**40.00**	**VI**
6.	Market Intelligence Agency	70.00	30.00	77.50	67.50	52.50	60.00	**59.58**	**II**

C1 = Circle 1, C6 = Circle 6

A perusal of Table 2.7, showed that infrastructural facilities like cooperative store at approachable distance for availability of fertilizer and agency for making market information available to the farmers on time received highest priority by the potato growers in the state with 66.25 and 59.58 of average infrastructure need index, respectively.

Government shop for supply of fungicide, insecticide at appropriate distance and provision for well-knit seed supply agency was the other infrastructure facilities were accorded high priority by the farmers with 58.75 and 54.17 need index. Potato storage related items like facility for cold storage and facility for country made storage godown, were the items that had low need index i.e. 46,25 and 40.00 respectively.

The high priority assigned to the establishment of cooperative storage for supply of fertilizer got support from the high training need on balanced use of chemical fertilizer like wise lower infrastructure need in storage related items also gets support from the lower training need (as per the respondents) area. One of the reason for according the storage related items like establishment of cold storage as well as country made storage structure a lower priority coupled with high need in establishing agency to market intelligence is that the potato growing tribal farmers finds a ready market of their potato produce within the state as well as markets in the neighboring north eastern states.

Owing to the fact that potato crop in the state is grown in summer, which is off-season for the neighbouring states in plain areas, thus, the need for storage is less in the state and whatever short term storage for 3-4 months is needed is to get a higher market price that can also be minimized if former have proper market intelligence of the domestic market as well as the potato market in the neighbouring states.

Constraints in the adoption of improved potato production, protection and storage practices as perceived by potato growing tribal farmers of Meghalaya

Constraints in the present book is operationalised as the reasons for non- adoption as well as problems encountered in adopting the scientific/ recommended potato production and post harvest handling practices. A total of ten numbers of constraints are identified related to potato production, plant protection, and storage, marketing as well farmers' own preferences. Ranking of these identified constraints has been done on the basis of number of respondents perceiving the constraint.

Table 2.8 : Constraints as Revealed by the Potato Farmers

Sl. No	*Constraints*	*Frequency (n = 120)*	*Percent-age*	*Rank*
1.	Rainfed conditions	97	80.84	II
2.	Non- availability of quality seed	97	80.84	II
3.	Fluctuating market price	102	85.00	I
4.	Preference for traditional practices of potato production	80	66.67	V
5.	Preferences for other better remunerative vegetable crops	78	65.00	VI
6.	Heavy disease incidence	88	73.34	IV
7.	Lack of scientific knowledge about scientific practice	93	77.34	III
8.	High cost of chemical inputs	71	59.17	VII
9.	Lack of proper storage	62	51.67	VIII
10.	Non- availability of shops for purchase of chemical inputs (Fert., insecticide, fungicide	80	66.67	V

The data summarized in the Table 2.8 revealed that 'fluctuating market price' has been perceived by majority of the respondents *i.e.*, 85 percent of the potato growers of the state and therefore, this constraint is accorded first rank. 'Non availability of quality seed' and 'rainfed agriculture' prevailing in this hilly state are both perceived by about 80.84 percent of the sampled tribal potato growers, and together they were assigned the second rank. About 77.50 and 73.34 percent of the sampled potato growers felt that lack of knowledge regarding scientific potato production practices and heavy disease incidence, respectively. These constraints were thus ranked third and fourth, respectively. Preference of potato growers towards traditional method of potato production and post harvest handling was the constraint that was perceived by 66.67 percent of the farmers. A similar percentage of the sampled tribal potato growers *i.e.*, 66.67 percent, perceived non-availability of shop for purchase of fertilizer, insecticide, fungicide etc. nearby the village as the problems in adoption of scientific potato production and post harvest handling practices. Thus both these constraints are accorded fifth rank. About 65 and 62.50 percent of the respondents perceived the problem of potato growers' preference for other better remunerative vegetable crops and high cost of inputs, respectively. High cost of chemical inputs was the constraint that is also perceived by 59.17 percent of the potato growers in the state. Lack of proper storage facilities in the state is revealed by a comparatively lesser percentage of respondents *i.e.*, 60 percent and thus it has been accorded eighth rank.

The reason of majority of the potato growers perceive the' fluctuating market price' as a constraint for adoption of recommended scientific potato production, protection and post harvest handling practices could be the fact that potato crop in the state, particularly in East and West khasi hill districts, is one of the most important cash crop that finds a readily available market. Thus, the returns from this crop are the major agricultural income of the farmers and the market price of potato thus in turn has a direct bearing on their total economic return.

The climatic conditions of the state are suitable for diseases like late blight to appear in an endemic form. In addition, viral diseases, brown rot, bacterial wilt are the other diseases common in the state. Thus heavy incidence of the diseases is also recognized by considerable percent of the respondents. Owing to this the potato produced in the State is not suitable for use as a quality seed. The seed thus is procured by the state agricultural department from outside agencies and is multiplied and distributed among the potato growers in the state. Hence, the formal potato seed distribution system is in operation in the state; however it is not enough to distribute seed as per the demand of the farmers in the state. Thus the non-availability of the quality seed is felt as a serious constraint by majority of the potato growers.

The probable reason for according a lower priority to the storage aspect could be the lower temperatures prevailing during the harvesting and the period after that leading to a lesser storage losses. In addition, the readily available market for sale of harvested potato within as well as outside the state reduces the need for scientific storage.

SWOT Analysis of Poatao Production in Meghalaya

SWOT is an acronym for **Strength, Weakness, Opportunity and Threat** associated with any subject under investigation. The four attributes are also called SWOT parameters. Strength is the competitive advantage the subject has for growth and development or in other words it reflects the basic asset with the subject. Weakness is the associated liabilities or the disadvantages for the growth and development. Opportunity reflects the potentials or ability to grow and achieve the specific objectives in a given situation. Threat is the situations or conditions that block the abilities to grow and develop.

SWOT analysis plays an important role in the understanding the problem associated with the subject under investigation, thus it helps in taking appropriate decision for its development in real operational environment. In an attempt to get a comprehension of the factors associated with the potato production in the state, SWOT analysis was carried out. The response on SWOT parameters was first taken using participatory discussion technique and the selected respondents were asked to rank the identified SWOT parameters. The parameters are being presented in the following subheads:-

Strengths of Potato Production in Meghalaya

Strengths are reflective of factors that have worked favorably so far. A total of six items are identified as strengths of potato cultivation in the state by the tribal farmers of the state. These identified/ perceived strengths are though not mutually exclusive holds importance in their own perspective. (Table 2.9)

Two Seasons of Potato Cultivation : The agro-climatic conditions of Meghalaya state especially of the hilly areas favours potato cultivation in two consecutive seasons i.e., the summer season (March – July) and autumn season (Aug – Nov). Though the production and productivity as well as area under potato is higher in the main summer season, the potato produce of autumn season fetches better market price for the farmers.

The two seasons in the state allows a higher potato production to cater to the domestic market in the state as well as allows export of potato to the neighboring states.

Potato Based Cropping Pattern : Potato is one of the most important crop in the state, especially in the hilly areas. In the hilly region, potato based cropping pattern Potato-Potato, Potato–Paddy, Potato–Cauliflower, Potato–Radish etc. are in practice. Farmers perceive the existence of such cropping pattern as the strength to Potato cultivation in the state. This potato based cropping system gets support from the semi-perishable nature of potato, that allows its short duration storage, transportation to long distances in the hilly state as well as timely marketing to fetch higher price. High domestic demand because of the well integration of potato in the dietary habits of the people of the state, also allows /provide support to this cropping pattern.

Local Method of Cultivation : Potato in Meghalaya is cultivated on an indigenous method of raised bed locally called as bun and nur method. This method has been evolved by the farmers through their own experimentations over the period of time. In this method raised beds made along the hill slope. These beds are made by clearing and burning the forest. This is a form of shifting or Jhum cultivation practiced in the state. This method has been adequately elaborated in the previous section. Farmers of the state therefore perceived this traditional method as strength of potato cultivation in the state. Farmers felt that this method of cultivation permits them to sustain even in the difficult conditions existing in the hilly state take the undulating topography, hilly terrain, farming on steep slopes coupled with heavy rainfall, as well as high disease incidence.

Export to the Neighboring States : Potato in the state is harvested during the months of May – July and then in Oct–Nov. These months corresponds with a period when Potato in the plain areas of neighboring state is in high demand. Growers in the state perceive the export of Potato from Meghalaya to the neighboring state as the biggest strength to the Potato cultivation in the state. Prevalence of Potato based cropping pattern coupled with two potato growing season, provide support for exporting of Potato in off season to the neighboring states of Assam,Tripura, etc.

Better Market Price : High demand of Potato in domestic market of state as well as in the neighbouring state in both the harvesting season, favours a high price of Potato. This is perceived as another major strength of Potato cultivation in Meghalaya by farmers of the state.

Weaknesses of Potato Cultivation in Meghalaya

Weaknesses are the negative characteristics and disadvantages associated with the situation, of potato cultivation in Meghalaya state. The major weaknesses as perceived by the farmers in the state are mentioned hereunder. (Table 2.10)

Rainfed Conditions : Like all other hilly regions of the country, agriculture in the state of Meghalaya is primarily rainfed with only 48 th. hectares of irrigated land in the state. Potato on the other hand is highly irrigation demanding crop. Though water requirement of potato crop is met in the state by heavy rain during the crop season. However crop many a times suffer from moisture stress, especially at the initial stage of growth in the main summer season.the rainfed condition hence is identified and judged as the most important weakness of potato cultivation in the state by the potato growers.

Lack of Scientific Know-How : Potato growing farmers in the state perceived lack of scientific methods of Potato cultivation among them as an important weakness of Potato production in the state. Evidently, as discussed earlier, farmers have perceived their local method of Potato cultivation as strength. However, non-refinement of traditional method in scientific line or in other words lack of knowledge of scientific methods pf potato cultivation leading towards low yield level was, therefore, felt as the serious weakness in potato cultivation. It is also found in the study that farmers in the state possess poor knowledge about the scientific methods of potato production like the improved potato varieties recommended for the state, the improved planting method,fertilizers dose and application method,insect-pest as well as disease management and storage.

Traditional Methods of Production : As potato cultivation in this tribal state of Meghalaya as discussed earlier is primarily carried out in traditional methods of bun-nur, with minimum inputs from the formal research system for the knowledge of use of mancozeb for management of late blight. It was interesting to note that though potato farmers perceived their traditional method of potato production as strength, they at the same time felt it as their weakness that ought to be refined and improved keeping in the local situation to improve the potato production and productivity in the state. Hence, farmers perceived it as one of the major weakness of potato production in Meghalaya.

High Disease Incidence : The climatic conditions of Meghalaya state are conducive for appearance of many diseases. Late blight is an important disease appearing regularly and in epidemic form in the state. The entire foliage of the potato crop is killed before it achieves full maturity leading to heavy damage to the potato crop in both the summer as well autumn season. In addition, brown rot, viral diseases, soft rot, early blight are other diseases appearing in the state. Farmers perceive the heavy incidence as weakness of potato cultivation in the state.

Heavy Rainfall : The name of the state 'Meghalaya' literally meaning- the abode of clouds is itself reflective of the extent of rainfall in the state.The rainfall in the state is very high and it varies from place to place in the state. According to the weather data of the year 1998 the rainfall varied from 2100mm in East Khasi hill district of the state to 16720 mm in Mawsynram region. In addition, the rainfall in the state is received over a long period of time starting from the month of April to month of September-October. The heavy rainfall was perceived as a weakness to potato cultivation in the state as it hampers all the associated activities of potato production, storage, transportation and marketing.

Opportunities in Potato Cultivation

Opportunities are the factors or situations that can

benefit, enhance or improve the potato cultivation in Meghalaya. Farmers' perceived six opportunities in the state associated with potato cultivation, those are being discussed in details in the following subheads. (Table 2.11)

Potato Rrocessing : Potato processing is mainly restricted to making potato chips, french fries etc., in the state. Potato chips that are sold in the markets of the state are either prepared from reputed brands or are produced in small scale processing units present in nearby Guwahati town or are prepared in the state itself.

Farmers' perceive value addition in form of chips, french fries etc., as a major opportunity for potato cultivation in the state. Small scale processing units based at village or family level could be set up to take advantage of the nearby markets for higher profit to the farmers from potato production in the state.

Better Linkage with Markets : Potato produced in the state finds market for itself within the state as well as in some of the neighboring north eastern states both as table and seed potatoes in the off season. This help potato grower in the state to fetch higher price for their produce. Establishing better linkages with the market of more number of neighboring north eastern states could lead to more export of potatoes from the state and also better returns to the farmers of the state.The potato growers in the state, therefore, perceived this as an important opportunity.

Large Scale Adoption of the Late Blight Resistant Cultivar : Kufri Jyoti and Kufri Megha are the improved high yielding cultivars being grown in the state. Kufri Megha though is highly resistant to the late blight disease, which is a major disease affecting the crop yield, is a poor bulker leading to a lesser yield. Kufri Jyoti on the other hand though is a fast bulker, but it has no more resistance against late blight disease, leading to a low yield. Kufri Giriraj is a new variety, which is a fast bulker as well as is highly resistant to the late blight disease, is a new introduction in the state that is fast becoming popular among farmers of the state. Therefore, in this background the potato growers in the state perceive large scale adoption of Kufri Giriraj and Kufri Girdhari cultivars as a major opportunity to the potato production in Meghalaya state.

Expansion of the Area : Potato is cultivated presently in about 20 th. hectares of land in the state of Meghalaya. The cultivation of potato in the state is primarily confined to the two districts viz., East and West Khasi Hills districts, located in the central plateau of the state. Horizontal expansion of potato producing could be done in the non- conventional potato producing pockets within and also outside these districts, wherever the climatic conditions are favorable. Farmers themselves perceive this as an opportunity for potato cultivation in Meghalaya state.

Large Scale Adoption of True Potato Seed (TPS) Technology: Majority of the potato farmers in the state continue to use the same seed stock for number of years without replacing seed, leading to decline in the potato production as well as productivity over period of time. The reason for this situation might be the poor availability of quality seed potato in the state or the lack of knowledge about scientific seed replacement rate to be practiced. Farmers in the state especially the farmers from the sampled villages as are aware of the advantages of TPS technology due to on- farm demonstrations, training etc., perceive the adoption of true potato seed technology as an opportunity for improving the quality seed availability thus enhancing the potato production and productivity.

Threats to Potato Cultivation **:** Threats are the factors, situation, that hinder potato cultivation in Meghalaya. A total of seven items are identified and ranked by the sampled potato growers, which are being discussed in details in the following paragraphs. (Table 2.12)

Fluctuating Market Price : Potato growers in the state primarily dispose their produce to the local market or to the local traders, who in turn sell the produce to different markets according to the demand. Thus the price of the produce is fixed according to the demand for potatoes in the market and the quantity of supply of potatoes by the farmers. Therefore, in the situation of glut, farmers are forced for distress sale at lower prices. It is because of this reason that farmers have started partially shifting to other vegetable crops in the state. Farmers perceived that the fluctuating market price is a major factor that can negatively affect potato cultivation ion Meghalaya.

Non-availability of Quality Seed : Quality seed is one of the most important inputs for profitable potato production in Meghalaya state. In the absence of adequate supply of quality potato seed, farmers in the state are forced to use the same seed continuously for number of years ultimately leading to a poor yield and reducing profits. Moreover, the favourable climatic conditions in the state for the diseases (bacterial, fungal, viral) and also the insect pest during the crop season, makes quality seed production difficult unless and until some very specific provisions are made. These provisions are difficult to put into practice by the farmers in their fields. Hence, this has been perceived as a serious threat to potato cultivation in the state by the farmers.

Competition from the Plain Areas : Potato is imported in Meghalaya from other states namely Uttar Pradesh, Bengal and Punjab etc for the major part of the year except for the months during harvesting. The early harvest of potato in the state is disposed off in the market during the last week of May to June. During this time, the produce face stiff competition from the potatoes imported from the plains, which is usually cheaper. Moreover, early harvesting of potatoes especially in low land is necessary as the same land is prepared for paddy crop in the crop sequence. Khasi farmers of the state thus during this harvesting time fetch a relatively lower price for their produce.

Late Blight Menace : Late blight is major disease appearing regularly every year and is severe in nature, causing heavy damage to the potato crop and thus reducing the yield of potato crop considerably. The disease appears usually around mid to last week of May in the state and the entire foliage is damaged within 2-3 weeks of appearance of the disease. The late blight therefore is perceived as a factor that can adversely affect potato cultivation in Meghalaya. Potato growing farmers in some of the areas where climatic factors are favourable, have devised a mechanism to reduce the effect of late blight menace by making adjustments in the potato planting time as well as in the cropping pattern. In these areas farmers take early crop from the period of February to June. Taking early crop helps potato beneficial to potato growers, as the potato crop is almost three months old when the blight appears, thus reducing the damage by the disease to some extent. However, in the autumn season crop late blight causes heavy damage. Secondly, by taking potato crop early, farmers' timely plant the next crop that in low land is usually paddy and in hill slopes other some vegetable crops. This is the reason that farmers in the state prefer spraying fungicide in the autumn potato crop rather than in the summer potato crop.

Competition with Other Crops : Vegetables crops like cabbage, tomato, radish, carrot, french beans, peas etc., find a ready market in the nearby cities and town, coupled with good market price. Potato in contrast, is comparatively a longer duration crop in the state (usually 5 months), requiring heavy inputs and most importantly the market price of the potato produce are not assured or stable. Thus, vegetable crops are now a days being preferred by farmers in comparison to the traditional potato crop especially by those farmers who are residing in the villages nearby cities. Competition from the other vegetable crops thus poses a serious threat to potato cultivation in the state.

Traditional Method of Potato Production Practices : Potato in the tribal and hilly state of Meghalaya is still being primarily cultivated using traditional practices. Improved scientific potato production practices are yet to reach to majority of the farmers in the state. Potato cultivation on the traditional lines is one of the major causes of poor yield of potato thus poor returns to the potato farmers in the state. Thus low adoption of scientific potato production practices negatively affecting the potato cultivation in Meghalaya state.

Brown Rot Incidence : Brown rot in potato crop appears in the summer as well s autumn potato crop in the state. the incidence of this disease varies from 10- 12 percent. The selected potato growers also perceive this disease as threat to the potato production in the state. However, this was judged as the least important threat by the potato growers.

Implications

The deliberations made in the preceding sections and

sub-heads have revealed that potato is the mainstay of agrarian economy in rural Meghalaya. This crop is well integrated into cropping pattern as well the dietary habits of people of Meghalaya. However, the production practices, in general, are traditional in nature. There is alarming gap in the level of adoption of recommended potato technologies. Also, farmers have a number of constraining factors limiting the adoption of the recommended technologies. In spite of the various inbuilt /inherent strength in potato production system in Meghalaya, the weakness and threats appear unsurmountable. Therefore, following implications could be suggested for researchers, extension and development personnel as well as policy makers.

Implications for Researchers

(*i*) Documentation of the traditional potato cultivation practices revealed farmers own method of land management by 'nur and bun' method. Therefore, this gives ample opportunity for the soil scientist to assess the physical attributes of soil, nutrients availability in it and other aspects of soil –plant dynamics in the above method of land preparation in comparison to the scientist /recommended method.

(*ii*) Similarly, traditional potato planting methods revealed two things. Firstly, there was high seed rate used because of the lower row-row distance (30-35 cm). Secondly, the placement of seed tuber, FYM and fertilizer mixture is exactly in the reverse order as those in recommended method. Therefore, assessment trials or experiments needs to be conducted to see performance of above traditional methods in comparison to the scientific recommendations both on-station and on-farm. An appropriate refined potato planting method could be evolved keeping in mind the farmers land structure and climatic conditions prevailing in the state.

(*iii*) Similarly, the existing HYV of potato in cultivation showed that K.Jyoti is the popular potato cultivar in the state despite it has lost its resistance to late blight may be because of the high bulking ability. Similarly K.Megha is still prevailing despite lower yield and poor shape (round) owing to its greater resistance against late blight thereby ensuring assured production. Though, the newly introduced Cv. K. Giriraj has both the qualities of high bulking rate, acceptable shape (oval) and high resistance against late blight, it has very poor keeping quality leading to high extent of rottage in store as well as in field. Therefore, attention is warranted for evolving more number HYVs of potato having all the attributes namely high /fast bulker, high resistance to late blight, oval shape and good keeping quality so that higher potato production and yield could be sustained.

(*iv*) It is a well known fact that Meghalaya is the hot spot of late blight. However, farmers traditional methods of late blight for summer potato revealed that they have adjusted the potato planting time especially in low land in such a way (through early planting) that they get sufficient bulking in potato (particularly for K.Jyoti) by the end of May and even if the crop dies by the early to mid June farmers get good yield. Therefore for summer crop, farmer's seldom spray their potato crop with fungicide. Hence, there is need to re –experiment the fungicide spray schedule for summer crop in Meghalaya for different land type and potato cultivars in order to assess and validate farmers traditional method of late blight management. Similarly for autumn crop, yet another spray schedule needs to be standardized and recommended.

(*v*) Similarly, as the existing situation in the state of Meghalaya showed heavy incidence of late blight and brown rot, therefore breeder seed production is discouraged. However, in order to meet the farmers' requirements of quality seed, agro-techniques for quality seed production ought to be standardized and disseminated among the potato farmers.

(*vi*) A number of traditional potato storage structures, and methods are documented in the state.These methods requires further investigation in terms of their feasibility, economics, physiological state of potatoes (table & seed) store there in and accordingly and improvement in their methods of potato storage.

Implications for Extension and Development Organizations / Departments.

(*i*) It has been found that there is very good availability of state extension institutions (both in terms of structure and function) at the grass root level with adequate manpower. However, findings showed that farmers are still continuing with their traditional methods of potato cultivation. Therefore, there is more responsibility on the part of state extension personnel to persuade and convince the potato farmers about the advantages of scientific potato cultivation by utilizing appropriate extension methodologies approaches *viz.*, on-farm demonstration, farmers meeting, crop exhibition, on-farm training etc.

(*ii*) The training need perception as showed by the potato farmers revealed the areas on improved potato production practices where farmers accorded their training needs. Therefore, sate extension functionaries ought to organize both on-campus and on-farm training programme involving both theoretical coverage and practical session in order to have greater impact on cognitive domain of potato farmers.

(*iii*) Similarly, several needs pertaining to infrastructure requirement in the state was revealed by the potato farmers.The related state development functionaries,therefore, plan and implement the infrastructure development schemes with special reference to potato in the state of Meghalaya.

Implications at Policy Level

(*i*) Assured market price for potato in the light of high cost of potato production in the hilly stae of Meghalaya is to be guaranteed. This will help in surmounting the farmers threat perception of fluctuating market price

(*ii*) The existing market intelligence ought to be further sharpened so as to make the market information readily available to potato farmers' well on time as well as in advance. This would protect farmers from the distress sale.

Table 2.9 : Major Strength of Potato Cultivation in the Study Area as Perceived by the Farmers

Sl No	*Strength*	*Rank* *C1*	*C2*	*C3*	*C4*	*C5*	*C6*	*Average Score*	*SD*	*CV*	*Final ranking*
1.	Two seasons for potato	VI	II	V	III	I	IV	**3.5** (72.00)	**1.87**	**53.43**	**III**
2.	Local methods of cultivation	III	III	III	V	IV	VI	**4.0** (65.00)	**1.26**	**31.62**	**IV**
3.	Well connected with city	IV	IV	IV	IV	V	V	**4.3** (64.00)	**0.52**	**12.00**	**V**
4.	Potato based cropping systems	I	V	I	II	VI	III	**3.00** (75.00)	**2.09**	**6.92**	**II**
5.	Better market price	V	VI	VI	VI	III	II	**4.67** (60.00)	**1.75**	**37.50**	**VI**
6.	Export to other states	II	I	II	I	II	I	**1.50** (80.00)	**0.55**	**36.50**	**I**

Figure in parentheses indicate percentages.

Table 2.10 : Major Weakness of Potato Cultivation Prevailing in the Study Area as Perceived by the Farmers

Sl No	Weakness	Rank						Average Score	SD	CV	Final ranking
		C1	C2	C3	C4	C5	C6				
1.	Heavy rainfall	II	III	V	V	V	III	**3.83**	**1.33**	**34.70**	**V**
2.	High disease incidence	V	II	III	I	IV	IV	**3.17**	**1.47**	**46.40**	**III**
3.	Rainfed condition	I	IV	I	II	I	I	**1.67**	**1.21**	**72.50**	**I**
4.	Traditional methods of production	III	V	II	IV	III	V	**3.67**	**1.21**	**33.00**	**IV**
5.	Lack of scientific know-how	IV	I	IV	III	II	II	**2.67**	**1.21**	**45.36**	**II**

Figure in parentheses indicate percentages.

Table 2.11 : Major Opportunities of Potato Cultivation Prevailing in the Study Area as Perceived by the Farmers

Sl No	Opportunities	Rank						Average Score	SD	CV	Final ranking
		C1	C2	C3	C4	C5	C6				
1.	Expansion of areas	IV	IV	II	V	II	IV	**3.50**	**1.22**	**35.00**	**IV**
2.	Better linkage with market	II	III	III	I	VI	III	**3.00**	**1.67**	**55.78**	**II**
3.	Potato processing	I	I	I	III	I	II	**1.50**	**0.84**	**55.78**	**I**
4.	Contact farming	VI	II	VI	VI	V	V	**5.00**	**1.55**	**30.98**	**VI**
5.	Large scale adoption of TPS	V	V	V	IV	IV	VI	**4.33**	**0.93**	**21.55**	**V**
6.	Large scale adoption of K.Giriraj	III	VI	IV	II	III	I	**3.17**	**1.72**	**54.33**	**III**

Figures in parentheses indicate percentages.

Table 2.12 : Major Threats of potato cultivation practices prevailing in the study area as perceived by the Farmers

Sl No	*Threats*	*Rank*						*Average Score*	*SD*	*CV*	*Final ranking*
		C1	*C2*	*C3*	*C4*	*C5*	*C6*				
1.	Non availability of quality seed	IV	IV	V	II	I	III	**3.17**	**1.47**	**46.43**	**II**
2.	Traditional cultivation practice	III	V	II	V	VII	VII	**4.83**	**2.04**	**42.26**	**VI**
3.	Competition with other commercial crops	VII	III	VI	IV	IV	IV	**4.67**	**1.50**	**32.24**	**V**
4.	Fluctuating market price	II	I	I	I	II	I	**1.33**	**0.52**	**38.83**	**I**
5.	Late blight menace	VI	II	IV	VI	III	V	**4.33**	**1.63**	**37.71**	**IV**
6.	Brown rot incidence	V	VI	IV	VII	V	VI	**5.50**	**1.05**	**19.67**	**VII**
7.	Competition from plains	I	VII	III	III	VI	II	**3.69**	**2.34**	**63.37**	**III**

Figures in parentheses indicate percentages.

CHAPTER 3

Farmers' Field Scenario of Potato Production in North Eastern States : Experiences of Assam & Nagaland

Comprehension of existing potato cultivation practices *ex-ante* is an important activity in order to decide the appropriate farm intervention. This is equally applicable for the dissemination of potato technologies. This includes understanding the farmers' traditional/indigenous production practices along-with the associated rationale, level of gap in the adoption of recommended technologies and perceived constraints in the adoption of recommended technologies. Also, right understanding of farmers' need perception related to production technologies, training, and infrastructure helps to select the appropriate technologies. Keeping the above facts in view, thorough survey of the three states in this region, namely, Assam and Nagaland was carried out on the above dimension and the findings so obtained are presented in following subheads :

Survey of Potato farmers of Assam

200 farmers from 6 villages of two district of Assam namely Nagaon and Barpeta were surveyed to collect some first hand information on potato cultivation. A multistage random sampling technique was utilized for selection of one block namely Mandia and four villages namely Bordoloni-2, Bordoloni-3, Paschim Jaharpam and Baharmura and 100 sample farmers (i.e. 25 from each village). Similarly, 100 farmers from Nagaon district were also sampled. The scientific potato production technologies were categorized into three heads namely potato production, potato protection and post harvest handling for comprehensible analysis and synthesis of

data. In this study, the farmers who followed the recommended practices were termed as adopter farmers and those who did not practice the recommended technologies were considered as non-adopter farmers. Similarly, based on the discussion with potato growers from the non-sample area in those villages, constraints were enumerated and classified into biophysical, socio-economic, technological and institutional constraints. The sample respondents were then asked to indicate the constraints experienced by them in adoption of scientific cultivation of potato in the state. Similarly, farmers' opinions for amelioration of those constraints were also ascertained on open-ended questions. The frequency of respondents indicating each of the constraints along with the suggestions was found out and expressed in percentage and finally ranking was done. The data were collected with the help of a semi-structured schedule coupled with group discussion in participatory environment. Simple statistics *viz.*, frequency and percent distribution were utilized to draw meaningful conclusions. The findings are presented for pooled data as below:

Traditional Practices of Potato Cultivation

Indigenous/traditional knowledge (IK) refers to the unique, traditional, local knowledge existing within and developed around the specific conditions of women and men indigenous to a particular geographic area (Grenier, 1998). IK does not refer to truth in its universal sense. Rather, it is knowledge, which is relevant to a given population in its most appropriate sense. IK is stored in peoples memories and activities and it is expressed in the form of stories, songs, folklore, proverbs, dances, myths, cultural values, believes, rituals, community laws, local language and taxonomy, agricultural practices, equipment, materials, plant sciences and animal breeds (Atte.1992). Acquisition of IK take place over a long period of time by local people through the accumulation of experiences, informal experiments and intimate understanding of the environment in a given culture that enables communities to survive.

IK has two powerful advantages over outside knowledge. It has little or no cost and is readily available. IK systems and technologies are found to be socially desirable, economically affordable, and sustainable and involve minimum risk to rural farmers and producers and above all, they are widely believed to conserve resources (Grenier, 1998). Keeping the above points in view, the present section of the book covers the farmer's indigenous/traditional practices of potato cultivation as well as the rationale behind them as potato is one of the important horticultural crops of the state of Assam and especially the Barpeta and Nagaon district.

Crop Production

Seed Management Practices : From the investigation it was seen that the farmers were bound to purchase the seed from the private seed supplier of their area as the government seed supply/quality seed are not available at planting time and at affordable price. Hence, the farmers mostly grew the varieties having no proper varietals identity sold by the private suppliers.The supplier does not sell the HYV as it is costly and are not affordable by the farmers. Again farmers expressed that they are unable to store their produce because of high rottage in store due to high temperature. Hence every year they procure the seed from the private seed supplier.

Potato Planting : The farmers go for cultivation of potato in the month of October-November as the climate is suitable for the potato crop. The farmers of the area reported that they mostly procure medium size tubers for planting and use recommended seed rate as the medium size tubers give good coverage. They rarely use big or small size tuber for planting as it is not so profitable to them to cover the area. Again the farmers expressed that they apply the FYM in very less quantity because of the non-availability of the FYM. The farmers follow the below-recommended spacing (i.e. 15cm × 15cm or 15cm × 20cm) because they do not have knowledge on proper spacing and its advantage. Most of the farmers make the drainage channel of 30-40 cm width and 30-cm depth to drain out the excess water during heavy rain.

From the investigation it was also clear that the potato farmers did not apply fertilizer either due to their financial problem and even if they apply they did not follow the recommended dose of fertilizer due to lack of proper knowledge on the application of balanced dose of fertilizer.

It is also found that the potato farmers of this area do not follow the scientific method of potato cultivation. As per the scientific method first furrow is made and then fertilizer is applied on the furrow. After that FYM is applied just above the fertilizer and then tuber is put there. After that it is covered with soil. Farmers, however, first apply the fertilizer and FYM in the soil after soil preparation and then they mix it in the soil by ploughing. After that they make furrow and put the tuber in furrow and then cover it with soil. This is their traditional practice and they do not have scientific knowledge on planting method.

Intercultural Operation : The potato farmers of this area expressed that they use rice straw or water hyacinth to cover the land, which act as mulching and also reduce the weed growth and thereby it helps to reduce the labour requirement for weeding and earthing up operation. Thus it reduces the cost of cultivation to some extent.

Crop Protection

Seed Treatment : The farmers informed that they mostly use medium size tubers and rarely the big size tuber which is to be cut. Hence, need for treatment of cut tuber with chemicals is not felt by them. Also they reported that they have no proper knowledge on seed treatment. The farmers growing the whole tubers also do not go for seed treatment because they are not aware of the advantages of it as well as do not have any knowledge on how to treat the seed and with which chemicals.

Disease Management : From the investigation it is seen that most of the farmers apply fungicide to control late blight disease to get higher yield with the limited knowledge on the control measures. They generally give 4-5 spray of Dithane M-45 @2.5 gm/litre at the interval of one week with the first spray at 30 days after planting. In the first spray they apply fungicide @ 100gm/45 litre of water, second @150 gm/90 litre of water and third @ 200 gm/ 100 litre of water. As the disease spread progresses they apply high concentration of chemical. Some of the farmers are not able to apply fungicide because of financial problem, though they are willing to control blight in their field.

According to the farmers the brown rot disease is not severe as compared to the late blight and hence they give little attention to this disease and if it appears in the field they just rouge out those plants and throw at distant places but do not use any chemical for this disease. The potato farmers also do not have proper knowledge on chemicals to be used for the control of brown rot disease. '

Insect Pest Management : The potato farmers of this area do not use chemicals against the insect pests because of the limited or no knowledge on the management of insect pest of potato as well as the high cost of chemicals. Infestation is higher during storage, they do not prefer to store potatoes and sell it just after harvest.

Extent of Adoption of Scientific Potato Cultivation Practices

Extent of Adoption of Scientific Potato Production Practices : Findings presented in Table 3.1 showed that no farmer of the studied area adopted practices like seed source utilized, use of quality seed for planting, spacing, weeding, earthing up operation and haulm cutting. On investigating the reasons, it was found that none of them purchased the seed potato from the reliable sources because of lack of its adequate availability and even if it was available the rate was so high that farmers could not afford to buy. Therefore, they were bound to purchase the seed potato from the private seed supplier of their locality every year which is usually of not good quality and specification. They purchase the seed tuber every year as they can not store their own potato for next season planting due to heavy rottage in the store and they can not even go for storing the potato in cold storage as it is far from the production site and usually not affordable. Again not a single farmer was found to adopt haulm cutting practice in potato crop which was mainly because of lack of knowledge. On the other hand, only 37% of the farmers were found to apply FYM in potato cultivation and the remaining 63% farmers did not go for FYM application due to the non-availability of the required amount of FYM at planting time. Again, it is evident from the table 6.1 that no farmers were found to adopt the practices like weeding and earthing up.The reason for non-adoption of practices like weeding and earthing up operation was that the farmers used locally and abundantly available rice straw/water hyacinth to cover their crop just after planting the seed in the furrow and hence, earthing up as well as weeding operation was not perceived necessary. On the other hand, cent percent non-adoption of the recommended spacing was because of lack of scientific knowledge. Cent percent of the farmers were found to adopt the recommended seed use pattern as well as proper seed size. Only 74% of the farmers were found to apply balanced dose of fertilizer as against the 26% non-adoption farmers which was because of lack of finances. Similarly, 92% of the farmers were found to apply recommended seed rate as against the 8% non-adopters reasoned due to lack of knowledge. The above findings get support from the study of Singh (1995) who reported that the technological gap to the extent of 34%, 42% and 44% were observed in potato cultivation among commercial, semi-commercial and non-commercial farmers of Kamrup district of Assam.

Extent of Adoption of Scientific Potato Protection Practices : In the area of plant protection, it is evident from the Table 3.2 that none of the sampled farmers was found to use fungicides to treat the seed tubers, for the control of brown rot/ bacterial wilt as well as use of insecticide to control the insect pests of potato. On investigating the reasons it was found that the farmers were lacking knowledge on use of any chemicals and were not aware of advantages thereof. However, 76% of the farmers were found to use fungicide for control of late blight and the reasons for 24% non-adoption were mainly because of lack of finance as well as lack of proper knowledge about mode of application, rate, dose, concentration, etc. Again, it is evident from the table 3.2 that 67% of the farmers practiced the important practice like rouging and 33% of the farmers did not go for rouging because of the unawareness of advantages as well as lack of knowledge among them. The above sets of constraints are probably true for other crops also and hence, Baroh *et al* (2004) identified them as the production constraints for rapeseed cultivation in Assam.

Table 3.1 : Frequency distribution of adopter and non-adopter farmers in the area of potato production

Sl. No.	*Recommended Practices*	*Total no. of pooled sample farmers (N = 200)*	
		Adopter farmers	*Non-adopter farmers*
1.	Seed sources utilized	0 (00.00)	200 (100.00)
2.	Use of quality seed	0 (00.00)	200 (100.00)
3.	Replacement of seed tuber in every 3-4 years	—	—
4.	Use of whole/ cut tuber as seed	200 (100.00)	0 (00.00)
5.	Seed rate used	200 (100.00)	0 (00.00)
6.	Spacing (Row to row and Plant to plant)	184 (92.00)	16 (08.00
7.	FYM application	0 (00.00)	200 (100.00)
8.	Balanced dose of fertilizer	74 (37.00)	126 (63.00)
9.	Application of Urea, SSP and MoP	148 (74.00)	52 (26.00)
10.	Weeding Operation	0 (00.00)	200 (100.00)
11.	Earthing Up	0 (00.00)	200 (100.00)
12.	Haulm cutting	0 (00.00)	200 (100.00)

Figures in the parentheses indicate percentage

Extent of Adoption of Scientific Post Harvest Handling Practices : Among post harvest handling practices it is evident from the Table 3.3 that cent percent of the farmers were found to adopt the practices like pre-storage drying and sorting of rottage/damaged tuber. Similarly 82% of the farmers were found to grade their tubers to fetch better market price and remaining 18% of the non-adopter farmers did not go for grading as most of them kept their produce for home consumption and were unaware with the importance of grading in fetching better price of their produce. On the other hand, none of the sampled farmers of the study area was found to go for scientific storage of potato after harvest which was due to lack of proper knowledge on scientific storage as well as lack of infrastructural facility.

Table 3.2 : Frequency Distribution of Adopter and Non-adopter Farmers in the Area of Plant Protection

Sl. No.	*Recommended Practices*	*Pooled sample of farmers (N = 200)*	
		Adopter farmers	*Non-adopter farmers*
1.	Treatment of seed tuber	0 (00.00)	200 (100.00)
2.	Use of fungicide for late blight control	152 (76.00)	48 (24.00)
3.	Scientific management of brown rot / bacterial wilt	0 (00.00)	200 (100.00)
4.	Use of insecticide	0 (00.00)	200 (100.00)
5.	Roguing	134 (67.00)	66 (33.00)

Figures in the parentheses indicate percentage.

Table 3.3 : Frequency Distribution of Adopter and Non-adopter Farmers with Regards to Post Harvest Handling

Sl. No.	*Recommended Practices*	*Total no. of pooled sample farmers (N = 200)*	
		Adopter farmers	*Non-adopter farmers*
1.	Pre-storage drying	200 (100.00)	0 (00.00)
2.	Soting of rottage/damaged tuber	200 (100.00)	0 (00.00)
3.	Grading of harvested tuber	164 (82.00)	36 (00.00)
4.	Scientific storage of potato	0 (00.00)	100 (100.0)

Figures in the parentheses indicate percentage.

Constraints as perceived by the potato growing farmers

Findings pertaining to the various constraints experienced by the potato farmers are delineated and discussed under the following sub-points :

Production Related Constraints : It is evident from the Table 3.4 that the lack of quality seeds was the most important constraints felt by 94 percent of the respondents which was probably due to lack of Government seed farms and village level seed producers. The similar finding was reported by Gogoi (2002).

Table 3.4 : Frequency distribution of the potato farmers based on the constraints experienced by them related to production aspects

(N = 200)

Production Related Constraints	*Frequency and Percentage*	*Ranking*
(*i*) Lack of quality seed	188 (94.00)	I
(*ii*) Lack of irrigation facility	098 (49.00)	IV
(*iii*) High cost of quality seed	162 (81.00)	II
(*iv*) Non availability of inputs like fertilizers, chemicals etc.	046 (23.00)	VI
(*v*) High cost of fertilizers and chemicals	080 (40.00)	V
(vi) Heavy incidence of disease and pest	124 (62.00)	III

High cost of quality seed was the second constraint reported by 81 percent of the farmers. This was because the quality seed is not being produced locally and they have to purchase it from distant places which add to the cost of seed. Heavy pest and disease incidence was ranked as the third important constraint as experienced by 62 percent of the respondents and this was primarily due to the growing of susceptible varieties by the farmers. Furthermore, most of the chemicals coming in the market were reported to be spurious and hence, had little effect on controlling disease as well aspest. The other bio-physical constraints reported were lack of irrigation facility (49%) followed by high cost of inputs (40%) and non availability of inputs like fertilizers, chemicals, etc. (23%).

Socio-Economic Constraints : It was observed from the Table 3.5 that lack of reasonable support price was felt to be the first important constraint by 91 percent of the respondents. Most of the respondent felt that the cost of production increased every year due to increased labour charges and input cost. It is quite understandable that farmers would have anticipated increased price for their produce when the cost of inputs go up.

Table 3.5 : Frequency distribution of the potato farmers based on the constraints experienced by them related to socio economic aspects

(N = 200)

Socio-economic related constraints	*Frequency and Percentage*	*Ranking*
(*i*) The scientific method of cultivation involves high cost and requires proper scientific knowledge	116 (58.00)	III
(*ii*) Price fluctuation in the market throughout the year	160 (80.00)	II
(*iii*) Lack of reasonable support price	182 (91.00)	I
(*iv*) High involvement of middleman	090 (45.00)	IV
(*v*) Scientific cultivation requires more labour and time.	062 (31.00)	V

Price fluctuation in the market throughout the year was the second constraint expressed by 80 percent of the respondents, which may be due to the lack of procurement facility by the Government. The similar finding was reported by Gogoi (2002). The third important constraint as expressed by 58 percent of the respondents was that the scientific method of cultivation involves high cost and requires proper scientific knowledge..The other socio-economic constraints were high involvement of middleman (45%) and more labour and time consuming scientific cultivation (31%).

Technological Constraints : It is evident from the Table 3.6 that the first and foremost technological constraint expressed by as high as 87 percent of the respondent was heavy rottage of tuber during storage which was due to higher temperature at the time of harvesting and storage. Non-availability of any suitable local method of storage was the second constraint expressed by 62 percent of the respondents and this aspect emanates researchable area to find suitable low-cost alternative method of storage. Similar finding was reported by Gogoi (2002). The other constraints were lack of proper knowledge on plant protection (45%) and lack of knowledge on HYV of potato for their region (34%).

Table 3.6 : Frequency Distribution of the Potato Farmers based on the Constraintsexperienced by them Related to Technological Aspects (N = 200)

Technology related constraints	*Frequency and Percentage*	*Ranking*
(*i*) Lack of knowledge on HYV of potato suitable to their region	068 (34.00)	IV
(*ii*) Lack of proper knowledge on plant protection	090 (45.00)	III
(*iii*) Heavy rottage of tuber in storage	174 (87.00)	I
(*iv*) No suitable local method of storage	124 (62.00)	II

Institutional Constraints : It may be observed from Table 3.7 that the most frequently perceived constraint as expressed by 71 percent of the respondent was lack of efficient marketing facilities at village level. Even though the Government or private owned marketing facilities were available at village level, they were not functioning effectively due to influence of middlemen. The producers had to pay varieties of incidental charges in unregulated as well as regulated markets. This considerably reduced the profit margin from the sale of the produce.

The yet another institutional constraint expressed by 65 percent of the respondent was non-availability of any cold storage facility nearby the production site/villages. Further discussions with the farmers revealed that only one cold storage with capacity as low as 50 MT exist to cater need of the potato growers of entire Barpeta district. By any standard, this is inadequate. The high rate of interest on private loan was the third major constraint expressed by 57 percent of respondents. Farmers were mainly going to the local money lenders due to easy procedure for obtaining loan in comparison to any financial institution. The other constraints revealed were complicated procedure for obtaining govt loan (48%), problem of transportation for marketing (43%), non-availability of credit facility (38%), insufficient amount of loan (31%) and no near by soil testing facility (05%).

Table 3.7 : Frequency Distribution of the Potato Farmers based on the Constraints Experienced by them Related to Institutional Aspects (N = 200)

Institution related constraints	*Frequency and Percentage*	*Ranking*
(*i*) No cold storage facility at the production site	130 (65.00)	II
(*ii*) Lack of efficient marketing facility at village level	142 (71.00)	I
(*iii*) Complicated procedure for obtaining Govt. loan	096 (48.00)	IV
(*iv*) In-sufficient amount of loan	062 (31.00)	VII
(*v*) High interest rate on pvt. loan	114 (57.00)	III
(*vi*) Problem of transportation for marketing	086 (43.00)	V
(*vii*) No soil testing facility available	010 (05.00)	VIII
(*viii*) Non availability of credit facility in the locality	076 (38.00)	VI

Farmers' Suggestions to Overcome the Problems of Potato Cultivation in the State : The farmers opined that provision of clod storage at the approachable distance from the village (89%) and arrangement for supply of quality seed adequately and timely by the govt. agency (81%) should be accorded highest priority (Table 3.8). Such perception would have emanated because of the fact that farmers need to purchase the seed tuber every year from the other states due to lack of cold storage facility. Therefore if a cold storage facility becomes available at the nearby distance, the farmers can easily store their produce for seed purpose for the next planting season which will be of better quality than the seed potato coming from outside of the state which has no assured varietal identity and purity.

Table 3.8 : Frequency Distribution of the Potato Farmers based on Their Suggestions to Ameliorate the Constraints Experienced by them (N = 200)

Sl. No	*Suggestions*	*Frequency and Percentage*	*Ranking*
1.	Provision of cold storage facility at approachable distance	178 (89.00)	I
2.	Arrangement for adequate and timely supply of quality seed by the govt	162 (81.00)	II
3.	Procurement of produce by government through minimum support price	154 (77.00)	III
4.	Training on scientific method of potato cultivation	136 (68.00)	IV
5.	Supply of production inputs like fertilizer, chemicals, etc on subsidized rate	116 (58.00)	V
6.	Developing low cost on-farm storage structure	094 (47.00)	VI
7.	Better transport facilities for efficient marketing	088 (44.00)	VII

Similarly the potato farmers gave third priority in provision of any kind of mechanism for procurement of produce by government agency through minimum support price (77%). The reason for this preference was that potato farmers did not get reasonable price of their produce due to the non-predictive agricultural marketing system with fluctuating market price. In addition, the potato farmers also suggested that they should be trained on scientific method of potato cultivation (68%) from time to time as majority of the farmers lacked proper knowledge on HYV seed, plant protection measures, etc which may be the non-physical factor attributing for the low production and productivity of potato in the state. The farmers gave 5th priority for supply of production inputs like fertilizer, chemical, etc on the subsidized rate (58%) which is often too costly to afford by most of the small and marginal farmers. In the same way, the farmers suggested for developing low cost on-farm storage structure (47%) so that they can store their produce with minimum cost in absence of cold storage facility. They also felt need for better transport facility (44%) for efficient marketing of potato. Due to bad road condition the transportation cost is usually too high and there is lack of easy and cheaper mode of alternate transport facility.

Survey of Potato farmers of Nagaland

Survey was also conducted among the potato farmers spread across six villages in two districts namely, Dimapur (plain production situation) and Kohima (Hills production situation) in the state of Nagaland. Data were collected from the 60 potato growing farmers from the sampled villages. Findings of the survey are presented as below:

Traditional potato production practices of the Farmers of Nagaland

The existing farmers' crop production practices of potato have been discussed below for each practice.

Seed Source Utilized by the Farmers : Findings showed that most of the farmers (i.e. 86.66%) of Dimapur district purchase their seed potato from the private seed supplier, whereas in Kohima district majority of the farmers (i.e. 66.66%) use their own seed for planting. The reason for purchasing seed from the private seed supplier may be that such suppliers are available in Dimapur area but in Kohima nobody in the area was found to use the Govt. seed.

Varieties Grown and Potato Planting Pattern : From both the sampled districts, majority of the farmers were found to grow local varieties which are 76.66% and 100% in Dimapur and Kohima, respectively. Only a few farmers (23.33%) of Dimapur district use variety Kufri Jyoti. However, all the farmers of both the districts did not replace their seed potato in a particular point of time to increase the productivity. Moreover, it was further revealed that cent percent of the farmers of both the districts used whole potato seed in planting. No farmer of these two districts was found to use the cut seed in planting and above 85% of the farmers of both Dimapur and Kohima district used medium size tubers and few farmers of both the districts used small and large size tubers.

Though in Dimapur district, cent percent of the farmers used recommended seed rate, in Kohima district they used higher seed rate than the recommended level. Also, in Dimapur, about 93.33% of the farmers applied fertilizer to the potato crop, while in Kohima only 53.33% of the farmers applied fertilizer. Cent percent of the farmers of Dimapur district used recommended dose of MOP and below recommended dose of DAP, while the farmers of Kohima district applied recommended dose of urea and DAP.

Traditional Plant Protection Practices of the Farmers

The existing plant protection practices of the farmers have been documented for each practice. It was clear from the study that no farmers of both the district treat the seeds with chemicals. Similarly, none of the sampled farmers from both the districts apply chemicals to control late blight, brown rot diseases and insect pests of potato.

Traditional Post Harvest Handling of Potato

It was revealed from the study that in Dimapur district most of the farmers (80%) sold their produce and approximately 50% of the farmers kept their produce for home consumption,

Table 3.10 : Existing Practices of Post Harvest Handling followed by the Sampled Farmers

Sl. No.	*Practices*	*Pattern of use*	*Dimapur district (n = 30)*		*Kohima district (n = 30)*	
			Frequency	*%age*	*Frequency*	*%age*
1.	Disposable pattern	Sold	24	80	15	50
		Home consumption	15	50	10	33.33
		kept for seed	3	10	16	53.33
2.	Pre storage	Yes	30	100	30	100
		No	0	0	0	0
3.	Storage done	Yes	24	80	26	86.66
		No	6	20	4	13.33
4.	Grading & sorting	Yes	30	100	30	100
		No	0	0	0	0

whereas in Kohima district 50% of the farmers sold their produce and slightly higher than that (53.33 %) kept their produce for seed. Further, cent percent of the farmers of both the districts sundried their seed before storage after proper grading and sorting (Table 3.10).

Technology Gap in Adoption of Scientific Cultivation Practices of Potato

The survey was carried out Dimapur and Kohima district of Nagaland. In Dimapur district, in the area of potato

Table 3.11 : Technology Gap in the Adoption of Recommended Potato Production Technologies

Sl. No.	*Practices*	*Dimapur district (n = 30)*		*Kohima district (n = 30)*	
		Gap%	*Frequency (%)*	*Gap%*	*Frequency (%)*
Potato Production Practices					
1.	Seed source	100	30 (100)	100	30 (100)
2.	Varieties grown	77	23 (76.66)	100	30 (100)
3.	Seed replacement	77	23 (76.66)	100	30 (100)
4.	Seed use pattern	0	30 (100)	0	30 (100)
5.	Treatment of cut tuber	100	30 (100)	100	30 (100)
6.	Seed size used	30	9 (30)	40	12 (40)
7.	Seed rate	0	30 (100)	100	30 (100)
8.	Planting time	0	30 (100)	0	30 (100)
9.	FYM application	100	30 (100)	0	30 (100)
10.	Planting distance	0	30 (100)	100	30 (100)
11.	Drainage channel	0	30 (100)	0	30 (100)
12.	Planting methods	100	30 (100)	100	30 (100)
13.	Fertilizer applied	6.66	2 (6.66)	46.66	14 (46.66)
14.	Urea applied	100	30 (100)	0	30 (100)
	SSP applied	100	30 (100)	100	30 (100)
	MOP applied	0	30 (100)	100	30 (100)
	DAP applied	100	30 (100)	0	30 (100)
15.	Intercultural operations	0	30 (100)	0	30 (100)

Improved Potato Protection Practices

1.	Chemical treatment of seed	100	30 (100)	100	30 (100)
2.	Management of late blight	100	30 (100)	100	30 (100)
3.	Management of brown rot	100	30 (100)	100	30 (100)
4.	Management of insect pests	100	30 (100)	100	30 (100)

Improved Post Harvest Handling Practices

1.	Disposable pattern	–	–	–	–
2.	Pre-storage drying	0	30 (100)	0	30 (100)
3.	Storage done	–	–	–	–
4.	Grading and sorting	0	30 (100)	0	30 (100)

agronomy, 100% gap was noticed in practices like seed source utilized treatment of cut tuber, FYM application, planting method, urea, SSP and DAP application. Like that way, 77% gap was noticed in practices like varieties grown and seed replacement. The gap was very low in practices like seed size used and fertilizer application which was 30% and 6.66%, respectively. There was no gap in practices like seed use pattern, seed rate, planting time, planting distance, drainage channel and intercultural operations (Table 3.11).

In Kohima district, in the area of potato agronomy, 100% gap was noticed in practices like seed source used, seed replacement, varieties grown, treatment of cut tubers, seed rate, planting distance, planting method, SSP application, MOP application, etc. In practices like seed size used and fertilizer application, the gap was 40% and 46.66%, respectively. No gap was noticed in practices like seed use pattern, planting time, FYM application, drainage channel, urea application, DAP application and intercultural operation.

In the area of plant protection, 100% gap was noticed in all the practices of plant protection in both the districts of the state. Similarly, in the area of post harvest handling, no gap was noticed in the practices like pre-storage drying and grading and sorting of potato in both the districts of the state.

Constraints in Potato Production in the State of Nagaland

Findings of Table 3.12 revealed that most frequently perceived constraints in both the districts of Nagaland were non-availability of seed from Govt. source (86.67%), lack of knowledge on fertilizers application (86.67%), lack of knowledge on the disease symptoms and their control measure (100%), no cold storage facility available nearby (83.33%) and no subsidy from the Govt (83.33%).

Table 3.12 : Constraints Analysis of Dimapur and Kohima District of Nagaland (n = 60)

Sl.No.	*Major constraints*	*Rank*	*Frequency*	*%age*
1	*2*	*3*	*4*	*5*
Pooled Data of Dimapur and Kohima District				
1.	**Problem related to seed supply**	I	52	86.67
	(*i*) Non-availability of seed from Govt. source.			
	(*ii*) Seed obtained from the market is of not good quality			
2.	**Problem related to scientific method of planting**			
	(*i*) Higher cost of production in improved method	I	60	100.00
3.	**Problem related to timely application of balanced fertilizers**			
	(*i*) The chemical fertilizers are costly	I	52	86.67
	(*ii*) Lack of knowledge on fertilizers application	II	48	80.00
4.	**Problem related to plant protection measures**			
	(*i*) Lack of knowledge on the disease symptoms and their control measure	I	60	100.00
5.	**Problem related to storage**			
	(*i*) High rottage in indigenous storage condition	I	60	100.00
	(*ii*) No cold storage facility available nearby	II	50	83.33

1	2	3	4	5
6.	**Problem related to marketing**			
	(*i*) No assured market to sell their produce	I	48	96.67
	(*ii*) Involvement of middleman is quite high	II	54	90.00
7.	**Problem related to finance**			
	(*i*) Non availability of credit facility at low interest	I	60	100
	(*ii*) No subsidy from the Govt.	II	50	83.33

CHAPTER 4

Seed Potato Systems in North Eastern Region of India

Seed is the critical input for potato production. Seed alone contributes to 38-40 percent of the total cost of potato production (Upadhya *et al,* 1995; Khatana *et al,* 1997). In North eastern region, among the several limiting factors attributing the poor potato productivity, non- availability of quality seed adequately and at reasonable price is the most important one. The fact that the entire region is handicapped for production of seed potato due to prevalence of virus vectors almost through out the crop season are responsible for degeneration of seed stocks (Gupta et al 2006) and non availability and high cost of quality seed that force farmers to use of the same seed stock for many generations are the associated factors for the low productivity figures of potatoes. According to an estimate about 98 percent of the farmers in the region uses own seed (Kadian, 2010) thus the formal system is suffice to meet the seed requirement for as low as only 2 percent of the potato area in the region. In additions, the climatic conditions of the north eastern hill region are conducive for many potato diseases like late blight, brown rot, bacterial wilt, thus making the seed production further difficult. Bringing quality seed from a distance of 2000 kms from north India incurs heavy transportation cost for the resource poor farmers of the region. The situation of low coverage of formal potato seed system amply highlights the importance of informal seed distribution system in the region. Thus, it is imperative to analyze the prevailing informal and formal seed system in the North Eastern region of India, the present study was conducted.

The analysis was done in the two major potato producing states i.e., Assam and Meghalaya, in the north eastern region of India. The states Assam and Meghalaya were selected purposively as they contribute maximum to the total potato production ie., 56 and 20 percent, as well as total acreage under potato crop i.e., 63 and 15 percent, respectively, as compared to other states in the region. Two major potato producing districts from Assam namely Barpeta and Nagaon and one major potato producing district from Meghalaya i.e., East Khasi Hills district were selected for the study. Further, two major potato producing blocks/circles were randomly sampled from the selected districts. Two villages and 10 farmers from each of the selected village were further selected to constitute the sample for the study. Thus, a total of 120 potato farmers were interviewed for the study. Besides, 20 traders from both the districts of Assam were also selected depending on their availability at the time of investigation to elicit their pattern of utilization of various sources of seed potato for marketing.

The variables included were sources utilized for seed potato, average market rate of seed and varietal preferences as perceived by the potato growers. Data were collected using semi-structured checklist through focused group discussion method. Simple statistics like frequency, percentages, average and rank were utilized to draw meaningful conclusions from the study.

Seed Sources Utilized by Potato Farmers

Seed sources utilized by farmers of Assam

Findings presented in the Table 4.1 showed that none of the selected farmers was keeping own produce for use as seed in the subsequent season and they were purchasing seed potato from the traders every year. Seed from state government of Assam was procured by as low as 2.50 percent of the sampled farmers.

In the state of Assam, potato is grown as the winter crop for which the harvesting starts from the month of January to the month of February. The temperature after the harvesting rises up sharply leading to heavy post harvest losses due to rottage as well as shrinkage. This makes on- farm long term storage for seed purpose nearly impossible. The problem is further aggravated in absence of cold storage facility near the production site. This compels farmers to dispose off the produce as early as possible after the harvest for table purpose. Thus for the seed potatoes the farmers purchased new seed every year from the potato traders.

Findings get support from the study by Pandit et al (2006) who also reported that farmers in Barpeta district of Assam state were not using quality seed of improved varieties for potato cultivations.

Seed sources utilized by farmers of Meghalaya

In East Khasi hill district of Meghalaya all the sampled farmers (100%), were observed using own seed for planting potato in the main season. In addition to use of own seed, 27.50 and 25.00 percent were observed procuring seed from fellow farmers from neighboring villages and fellow farmers of same village, respectively for raising potato crop. Data presented in Table 4.1 show that about 15 percent of the farmers procured seed either from the state government directly or from state government registered potato seed growers from adjoining areas. In addition, seed from local traders was also reported to be procured by 12.50 percent of the sampled potato farmers.

The reason for practicing long term storage of seed as well as table potato in Meghalaya could be because of the prevalence of cooler temperature after the harvest season and farmers preference for potato varieties with a better shelf life.

Average Rate of Procurement of Seed Potato

Average Rate of Purchase of Seed Potato in Assam

Data presented in Table 4.2 indicate that farmers purchased seed potato between the months of September to November from different available sources. The varieties grown in the selected two district of Assam were Kufri Chandramukhi, Bengal Jyoti, Potatoes from Shillong and Bhutan. Kufri Chandramukhi and Kufri Jyoti were imported as seed from neighboring West Bengal state while harvest of summer potato crop from Shillong was imported as seed by the farmers of

Table 4.1 : Frequency Distribution of Seed Source Utilized by Potato Farmers in Assam and Meghalaya

Sl Source	*Assam*			*Meghalaya*		
	Nagoan (n = 40)	*Barpeta (n = 40)*	*Total (N = 40)*	*Mawphlang (n = 40)*	*Smit (n = 40)*	*Total (M = 40)*
1. Own seed	00(00.00)	00(00.00)	00(00.00)	20(100.00)	20(100.00)	40(100.00)
2. Fellow farmers of same village	-	-	-	04(20.00)	06(30.00)	10(25.00)
3. Fellow farmers from neighboring villages	-	-	-	09(45.00)	02(10.00)	11(27.50)
4. Seed from trader	40(100.00)	40(100.00)	80(100.00)	05(25.00)	00(00.00)	05(12.50)
5. Seed from state govt. or through registered growers	02(05.00)	00(00.00)	02(2.50)	01(05.00)	05(25.00)	06(15.00)

Figures in parentheses indicate percentages

Assam state. In addition, the potato seed was also imported from neighboring country like Bhutan. Thus majority of the demand for seed potato in the selected districts of Assam state was met through the imports from the neighboring states. Lack of availability of adequate storage infrastructure in the state also promoted the import of seed potatoes from neighbouring states as the country like Bhutan.

Table 4.2 : Average Rate of Purchase of Seed Potato by the Sampled Households in Assam

(N = 120)

District	*Potato varieties purchased by farmers*	*Rate of purchase*	*Month of seed/qt*	*Preferred Seed size purchase*
Barpeta	Bengal Jyoti	400-700	Sep-Nov	Small-Medium
	Kufri Chandramukhi	350-600	Sep-Nov	Small-Medium
	Potato from Bhutan	800-950	Sep-Nov	Small-Medium
Nagaon	Kufri Chandramukhi	400- 800	Sep-Nov	Small-Medium
	Bengal Jyoti	400- 800	Sep-Nov	Small-Medium
	Potato from Shillong	300-700	Sep-Nov	Small-Medium

The seed potato imported from Bhutan was purchased by farmers at the premium price (Rs 800-950 per quintal) while potato from Shillong fetched lesser price (Rs 300-700 per quintal).

It is further evident form the same table that in Barpeta district of the state, the rate at farmers purchased the seed of Kufri Chandramukhi and Bengal Jyoti varieties was relatively lower as compared to Nagaon district. This could be attributed to the proximity of Barpeta district to the West Bengal state from where these varieties for seed purpose were purchased easily. In addition, potato from Shillong was observed to have a market as seed in both the districts of Assam, while potato from Bhutan was utilized as seed in Barpeta district only.

This again could be attributed to the vicinity of the seed source to these districts.

The seed potato market was thus observed to be spread across different states of the region. West Bengal, Assam, Meghalaya were the major states catering to the seed demand of the potatao farmers of Assam.

Average Rate of Purchase of Seed Potato in Meghalaya

Potato farmers of East Khasi Hill district of Meghalaya were observed to purchase seed potato during the month of September to October (Table 4.3) for planting for the summer crop i.e. in the month of February and March. The reason for the delayed purchase which is more than three months after the harvesting of summer crop was mainly because of the prevailing perception among farmers that by this time the diseased seed tubers rots completely which helps to prevent the losses to the farmers.

Table 4.3 : Sources Utilized and Average Rate of Purchase of Seed Potato in Meghalaya

Sl. no	*Seed source*	*Month of purchase*	*Size of seed*	*Rate of purchase (Rs/q)*	
				Mawphlang Block	*Smit Block*
1.	Fellow farmers from the same village	September-October	Small	600	450
2.	Fellow farmers from the neighboring village	September-October	Small	600	450
3.	Farmers from neighboring districts	September-October	Medium Small	1000 750	- -
4.	Seed from Potato traders	September-October	Small	500	500
5.	Meghalaya potato experimental state farm	October	Medium	550	550

It could also be observed from the same table that the rate of purchase of seed potato was found to be much higher in Mawphlang block (Rs 600/q to 1000/q) as compared to the Smit block (Rs 450/ q) of the district. This could be attributed to the fact that Smit block of East Khasi hill district was covered under "Registered growers scheme' wherein state department of Horticulture multiply the potato seed which is primarily Kufri Jyoti and Kufri Giriraj varieties at farmers' fields and purchased it back for further distribution in thestate. The registered potato growers supply the surplus seed potato to the other farmers of the block informally at lower price.

Farmers from Mawphlang block were observed paying a higher seed rate for the seed bought from the fellow farmers from neighboring pockets of west khasi hills district. The reason for this could be the fact that these pockets are regarded as good quality seed source among the farmers. Seeds from traders and from the state potato experimental farm were available to farmers of the district at much lower rates i.e., Rs 500/- and Rs 550/q, respectively.

Thus in Meghalaya state, the demand of seed potato was primarily met from within the state and major seed source utilized by farmers was fellow farmers.

Varietal Preferences of Potato Farmers

Varietal Preferences of Potato Farers in Assam State

Bengal Jyoti variety of potato, a local selection from Kufri Jyoti variety by the farmers of West Bengal, was observed to the most preferred and imported seed potato variety by the sampled farmers of Assam (Table 4.4). Low seed price, satisfactory production (140q/ha), good tuber shape and size were cited as the reasons by the potato growers for their preference for Bengal Jyoti variety and Kufri Chandramukhi variety. It is important to mention here that potato traders were the major source of seed to farmers in the state. Seed potato from Bhutan though was perceived to give higher production, was preferred after Bengal Jyoti and Kufri Chandramukhi in Barpeta district of Assam, mainly because of the high seed price and its late arrival in planting season.

Potato arriving from Shillong was least preferred for the seed purpose in both the districts of Assam owing to its low production potential. The reason for this could be the fact that potato arriving from Shillong is generally small sized as well as usually mixture of different varieties of summer potato crop, which fail to give high yields in Assam. Moreover, the physiological stage of the seed potato produced from Meghalaya may not be biologically compatible Assam climate condition and cultivation time.

Table 4.4 : Varietal Preference of Potato Farmers in Assam and Reasons Thereof

District	*Variety*	*Preference*	*Reasons*
Barpeta	Bengal Jyoti	I	Low seed priceSatisfactory production (140q/ha) Good tuber shape and size
	Bhutia	II	Very good production (200q/ha) Highly remunerative seed price
	K. Chandramukhi	III	Satisfactory production (120q/ha)
Nagaon	K. Chandramukhi	I	Uniform shape and size of tuber Higher yield (160q/ha)
	Bengal Jyoti	I	More yield (160q/ha)
	Shillong seed potatoes	II	Lower yield level (130q/ha)

Varietal Preferences of Potato Farmers in Meghalaya State

The preferences of Meghalaya farmers for potato varieties varied according to their use. Therefore, it was essential to ascertain their varietal preferences separately for commercial and table purposes. For commercial purpose farmers of East Khasi Hills district of Meghalaya preferred Kufri Jyoti in both the blocks (Table 4.5). Satisfactory productivity, good tuber shape and size, storability, good market price and better taste were the reasons for its preference. Better storability of Kufri Jyoti was another important determinant for its preference as seed potato in Meghalaya under traditional storage system even for longer periods. This was followed by the preference for Kufri Giriraj and Kufri Megha. The reason for preferring Kufri Giriraj variety was the high production level and larger tuber size that fetched a better market price due to consumer preference. However, the least preferred variety among the farmers for commercial purpose was the local varieties like *Lah Sow, Lah Taret, Lah syntiew, etc.*

Table 4.5 : Varietal Preference of Potato as Perceived by the Farmers in Meghalaya and Reasons Thereof

Area	*Preferred Potato Variety*	*Preference for commercial purpose*	*Preference for home consumption*	*Reasons for preferences*
1	*2*	*3*	*4*	*5*
Mawphlang block	Kufri Jyoti	I	III	Satisfactory production Good tuber shape and size. Better storability Good market price
	Kufri Giriraj	II	IV	Higher production Large tuber size Better market price
	Kufri Megha	III	II	Higher yield than local Yield stability Good tuber shape and size Good Storability Better cooking quality and taste
	Local varieties	IV	I	Better in taste Low yielder
Smit block	Kufri Jyoti	I	III	Satisfactory production Good tuber shape and size Better storability Good market price

(Contd...)

1	2	3	4	5
	Kufri Giriraj	II	IV	Higher yield level Better market price
	Kufri Megha	III	II	Yield higher than local Late blight resistance Good tuber shape and size Good storability Better cooking quality and taste
	Local varieties	IV	I	Better taste and cooking quality Low yielder

In contrast, local varieties were observed to be the most preferred for home consumption, followed by Kufri Megha. Kufri Jyoti followed by Kufri Giriraj was least preferred varieties for home consumption in both the blocks of the districts. The reasons for poor preference for Kufri Giriraj was its poor keeping quality. On the other hand, Kufri Megha was the preferred owing to its better taste and better storability.

Sources of seed potato utilized by potato traders in Assam

A sample of 20 wholesalers from Assam was also selected for the study to find the different sources of seed potatoes utilized by them. The sampled potato wholesalers of Barpeta and Nagaon districts in Assam were observed importing potato from potato markets of adjoining states like West Bengal and Meghalaya, as well as the neighboring country, Bhutan for meeting the potato seed demand in Assam (Table 4.6). In addition, traders from Guwahati, capital city of Assam imported seed from different states like, Uttar Pradesh, Bihar, Punjab for further sale in North eastern states. Though many sources were utilized by traders for purchase of potato, the quantity of potato purchased from a particular source was determined by the prevailing market price.

It could be seen that (Table 4.6) the average rate of purchase and sale of seed was higher for Kufri Chandramukhi and Bengal Jyoti in Nagaon district as compared to the Barpeta

Table 4.6 : Sources of Seed Potato Utilized by Potato Traders in Assam and Rate of Purchase and Sale.

(N = 20)

Sl. No.	*Source*	*Barpeta (n = 10)*			*Nagaon (n = 10)*		
		Freq. (%)	*Rate of purchase (Rs/q)*	*Rate of sale (Rs/q)*	*Freq.(%)*	*Rate of purchase (Rs/q)*	*Rate of sale (Rs/q)*
1.	West Bengal markets	10 (100.00)	300-350	350-400 (50)	10 (100.00)	300-450 (50-100)	400-500
2.	traders from Guwahati	10 (100.00)	275-350	300-400 (25-50)	10 (100.00)	275-350	300-400 (25-50)
3.	Local potato from Assam	10 (100.00)	230-300	260-400 (30-100)	10 (100.00)	300-450	350-500 (50)
4.	Shillong, Meghalaya	04 (37.50)	300-375	360-425 (50-60)	03 (25.00)	250-300	300-360 (50-60)

Figures in parentheses indicate profit margin (Rs/q)

district of Assam, while it was samel in case the potato was imported from Shillong. The reason for this could be the low transportation cost involved in importing seed potatoes from West Bengal which is adjoining to Barpeta district as compared to Nagaon. It was also observed that traders from Nagaon district of Assam did not prefer to sell Kufri Jyoti and potato from Bhutan for seed purpose. In addition, these were to be purchased as well as sold at a higher rate as compared to other varieties.

Conclusion

The findings as elaborated in the above paragraphs clearly indicated that the informal system of seed potato marketing was primarily prevalent for meeting the seed demand of the potato farmers of north eastern region. Assam despite having a large area under potato immensely depends on the potato traders for supply of the seeds from neighboring states. The lack of adequate cold storage structures or low cost improvised on-farm storage structures for potato storage is the reasons for dependency of farmers on the traders. The situation demands suitable interventions related to setting up of seed multiplication farms and appropriate storage structures in the state. Similalrly, in Meghalaya, the limited operation of formal seed potato distribution system further needs to be strengthened for a larger coverage of potato farmers.

CHAPTER 5

Marketing of Potato in North Eastern Region : A Micro-Level Analysis

Marketing is defined as the commercial functions involved in transferring goods from producer to consumer. Marketing assumes special significance with regard to profitability of any agriculture commodity in a given agriculture production situation. This is more so with regard to perishable and semi perishable as well as bulky agricultural commodities.

Potato is an important crop in the North Eastern Region in India especially the hilly tracts, where the crop is grown under rain fed conditions. The crop is grown throughout the year in one or the other part of the North Eastern region contributing about 10 percent of the total area under potato in the country (Gupta *et al*, 2004). The said crop forms an important part of prevailing cropping systems (Kumar, et al 2006) as well as the dietary food habits of the people of the region. The semi perishability and bulkiness are the innate characteristic of potato that causes problems in its marketing. Marketing of potato in North eastern states of India is further constrained by hilly topography, that comprise about 70 percent of the total area (Kadian *et al*, 2010), which limits its movement which ultimately affects the resource poor farmers of the region. Marketing and storage were also identified as important weakness of potato production in India (Anonymous, 2007). Saikia (2001) also highlighted that the facilities of storage, processing and marketing are particularly deficient for perishable commodities as technological constraint for agricultural development in north eastern region of India. Owing to the highlighted weaknesses, the rural marketing in the region is confined primarily to unorganized sector with domination of private traders. The present chapter, therefore, examines the potato production and yield, marketable surplus, marketing pattern, the marketing channels utilized, and the preferred sources of table potato in the north eastern region of India.

Study Area

The study was conducted in two major potato growing states of the North eastern region of India, i.e., Assam and Meghalaya. For the study, two major potato growing districts i.e., Barpeta and Nagaon districts were selected from Assam state whereas East Khasi hills district was selected from the state of Meghalaya.Two potato growing circle/ blocks were selected randomly from each of the selected district for conducting the study. Random selection of two villages from each of the selected block/circle was done and 10 potato farmers were selected randomly from each of the village. The total sample size for the present study comprised of 120 potato growing farmers. In addition, 24 potato traders (8 from each selected district) constituted the present sample. The primary data was collected using a pretested interview schedule.

Major Findings

Potato area production, consumption and disposal in Assam and Meghalaya

The average land holding size among the sampled farmers in Barpeta and Nagaon districts of Assam was found to be 1.36 and 1.67 ha, wherein average land under potato cultivation was observed to be 0.63 and 0.64 ha (46 % total land), respectively. The average potato production among the respondents in Barpeta district of Assam was found to be about 100 quintal per season with average productivity of 158 quintals/ha, whereas in Nagaon district the average potato production was found to be 81 quintals with average productivity of 128 quintals per hectare (Table 5.1).

It could also be observed from the same table that the annual potato consumption in the households of respondents in Barpeta and Nagaon districts of Assam was 4.65 and 4.12 quintals which was about 4.65 and 5 percent of the total potato production at the household level, respectively. The average annual potato disposal was found to be 86 quintals in the selected districts of Assam. Average annual potato disposal was found to be 95 quintals per household in Barpeta and 77 quintals in Nagaon district of Assam.

Table 5.1 : Production, consumption and disposal scenario of potato among the sample farmers in Assam

District	*Sample Blocks*	*Av. Land Holding (ha)*	*Av. Area under potato (ha)*	*Av. Yield of Potato Obtained (q/ha)*	*Av. Prodin of Potato Obtained (q)*	*Av. Annual Con-sumption (q/house-hold)*	*Av. Annual Potato Disposal (q/house-hold)*
Barpeta	Mandia	1.50	0.63	160.09	100.86	5.10	95.76
	Bhavanipur	1.22	0.64	155.69	99.64	4.20	95.44
	Average	1.36	0.63	158.19	100.25	4.65	95.60
Nagaon	Dhing	1.09	0.58	130.08	75.45	3.64	71.81
	Rupahi	1.67	0.69	126.46	87.26	4.60	82.66
	Average	1.38	0.64	128.59	81.35	4.12	77.23
Grand Average		**1.37**	**0.63**	**143.08**	**90.80**	**4.38**	**86.42**

Among the selected districts of Assam, potato productivity, production, consumption as well as the disposal was found to be higher in Barpeta district.

Data presented in Table 5.2 indicated that the average land holding size in East Khasi Hill district of Meghalaya was 1.35 hectares, while the average area under potato was 1.24 ha (92%). The average potato production and yield was observed to be 137 quintals and 109 q/ha in the selected districts, respectively. The selected farmers were observed disposing on an average 130 quintals of potatoes and consuming about 6.85 q (5%) of potatoes annually. Among the selected two blocks, the average land holding size, the average land under potato, average yield, production, consumption as well as disposal was found to be higher in Mawphlang block in comparison to Smit block of Meghalaya.

From the results presented in the above paragraphs reflected the importance of potato in the cropping as well as the dietary habits of people in the North eastern region of India. The figures also reflect on the importance of potato crop to the farm economy of the selected districts of the state. The average potato productivity in Assam was found to be higher than in Meghalaya; however, the average household consumption of potato per year was much higher in Meghalaya than in Assam.

Table 5.2 : Production, consumption and disposal scenario of potato among the sample farmers in Meghalaya

Sample Blocks	*Av. Land Holding (ha)*	*Av. Potato Area (ha)*	*Av. Yield of Potato (q/ha)*	*Av. Prodn. of Potato (q)*	*Av. Annual Potato Consumption (q/ Household)*	*Av. Annual Potato Disposal (q/ Household)*
Mawphlang	1.59	1.43	115.6	165.31	7.10	158.21
Smit	1.11	1.06	102.4	108.54	6.60	101.94
Average	1.35	1.24	109	136.92	6.85	130.07

Potato disposal pattern over period of time in Assam and Meghalaya

It could be seen from the Table 5.3 that the respondents disposed off the total quantity of small and mixed potatoes (unsorted) immediately after the harvest in both the districts of Assam state. In addition, maximum quantity of medium sized potatoes (28.71quintals) and large sized potatoes (11.05quintals) were also disposed off by the farmers of both the districts immediately after harvest.

Further, about 15-45 days after harvest, sampled farmers disposed average 14.96 and 6.59 quintals of medium and large size potato, respectively which was further reduced to 9.06 and 3.15 quintals of medium and small size potato, respectively after 45-60 days of harvest. Thus, maximum quantity of potatoes harvested were disposed off immediately after harvest followed by after storage for 15-45 days and least quantity was disposed after 45-60 days of storage (Table 5.3).

Table 5.3 : Disposal pattern of table potato in Assam

Distt.	*Small*		*Medium*		*Large*		*Mixed*	
	Quan. Sold (q)	*Rate (Rs./ qt)*	*Quan Sold (q)*	*Rate (Rs/ qt)*	*Quan-tity Sold (q)*	*Rate (Rs/ (qt)*	*Quan. Sold (q)*	*Rate (Rs/ qt)*
Immediately after harvest								
Barpeta	10.68	113.66	30.00	213.55	12.00	267.24	7.38	235.00
Nagaon	9.91	109.48	27.43	379.50	10.10	442.23	4.86	359.16
Average	10.29	111.57	28.71	296.53	11.05	354.74	6.09	297.08
15-45 days after harvest								
Barpeta	–	–	14.02	262.08	6.10	338.91	–	–
Nagaon	–	–	15.90	428.55	7.09	486.84	–	–
Average	–	–	14.96	345.32	6.59	412.87	–	–
45-60 days after harvest								
Barpeta	–	–	9 06	366.67	4.00	417.85	–	–
Nagaon	–	–	–	–	2.30	540.83	–	–
Average	–	–	9.06	366.67	3.15	479.34	–	–

It could also be observed from the same table that with the increase in the duration after harvesting to 45-60 days, the price of potatoes kept on increasing, however, the quantity of potatoes disposed off by farmers reduced. This clearly indicated that farmers did not prefer storage of potatoes for long after harvest. The reason for this preference could be the poor storage facilities available to the farmers as well as the high temperature in the post harvest months of march- April prevailing in the state of Assam.

It could be seen that the quantity of small sized potatoes that was disposed off immediately after harvest was smaller than that of medium or large sized potatoes. This result could be attributed to the reason that storage of small sized potatoes cause considerable shrinkage making them unfit for marketing, therefore the small sized potatoes were predominantly used for home consumption in the farm households in Assam.

The data on potato disposal pattern among farmers of Meghalaya showed that a substantial amount of potatoes were disposed off immediately after harvest i.e, about 6.50, 5.03, 2.23, 1.26 quintals of small, medium, large and extra large potatoes, respectively. The price of potatoes immediately after was observed to vary from Rs 150 to 750/quintal with the variation in size of potatoes from small to extra large. (Table 5.4) It was also observed that potatoes harvested from low land in the state were predominantly sold immediately after harvest as farmers perceive them to be more diseases prone thus more prone to rottage in store.

Table 5.4 : Disposal pattern of potato harvested in Meghalaya

Particulars	*Small*		*Medium*		*Large*		*Extra Large*	
	Quan. Sold (q)	*Rate (Rs./ qt)*	*Quan Sold (q)*	*Rate (Rs/ qt)*	*Quantity Sold (q)*	*Rate (Rs/ (qt)*	*Quan. Sold (q)*	*Rate (Rs/ qt)*
Immediately after harvest(June-July)								
Mawphlang	10.22	150	6.84	350	3.99	575	2.00	750
Smit	2.79	150	3.21	350	0.47	575	0.52	750
Average	**6.50**	**150**	**5.03**	**350**	**2.23**	**575**	**1.26**	**750**
Within 45 days after harvest								
Mawphlang	9.21	300	16.61	550	8.64	650	1.10	800
Smit	11.42	300	14.39	550	6.15	650	1.00	800
Average	**10.32**	**300**	**15.50**	**550**	**7.39**	**650**	**1.05**	**800**
45-75 days after harvest								
Mawphlang	4.7	300	28.62	700	12.12	800	1.40	850
Smit	2.83	300	13.42	700	8.65	800	1.10	850
Average	**3.76**	**300**	**21.02**	**700**	**10.38**	**800**	**1.25**	**850**
75-105 days after harvest								
Mawphlang	1.13	150	4.15	300	2.24	400	–	–
Smit	1.00	150	3.01	300	0.80	400	–	–
Average	**1.06**	**150**	**3.58**	**300**	**1.52**	**400**	–	–

The quantity of potatoes disposed of all sizes *i.e.*, from small to extra large, increased with increase in the storage time from immediately after harvest to 45-75 days after harvest in unison with the increase in the market price (Table 4). However, after 75 days of harvesting, the market price for all size of potatoes was observed to reduce substantially. In addition, a maximum quantity of small sized potatoes (16.82q) was observed to be disposed off till 45 days after harvesting while medium (21 quintals) and large sized (10.38quintals) tubers were preferred for disposal during 45-75 days after harvest. The reason for this difference could be the change in market price for different sizes of tubers over period of time after harvest. The prevailing low temperature during harvesting (June-July months) and post harvest months in this hilly state, favours on farm storage of harvested potatoes for a longer duration of time as a profit maximization strategy after the initial rush of freshly harvested potatoes in market. Kadian etal (2010) also reported that potato farmers in Meghalaya state practiced staggered harvesting to avoid gluts on the markets and to get better prices for their produce

Marketing channels utilized by farmers for disposal of table potato in Assam and Meghalaya state

Data presented in Table 5.5 shows that 'Assemblers coming to the village' and 'local traders operating in the villages were observed to be the channels utilized by majority of the farmers in both the districts of Assam for disposing the medium, large and mixed sized(unsorted) potatoes after harvest. Small sized potato in the sampled districts was disposed to the local traders in the village by as low as 22.50 percent of the farmers. The reason behind this could be the fact that as the small sized tubers as fetched minimum price in the markets, they were utilized for home consumption in majority of the households in the districts.

Medium and large sized potatoes were observed tc be sold directly to city markets or to the whole sellers coming to the villages to collect the produce under contract by a significant percent (17.5-25%) of the selected potato farmers in Assam. Under the said contract, the whole sellers provided potato seed to the farmers on credit at planting time, who in turn sells the harvested produce to the whole seller at relatively lower price.

From the results presented above, it could be inferred that majority of the farmers prefer to dispose potato at farm itself and only a few prefer to dispose it in market. The reason for this could be the poor transportation facilities coupled with poor road conditions that made the transportation cost high for the farmers.

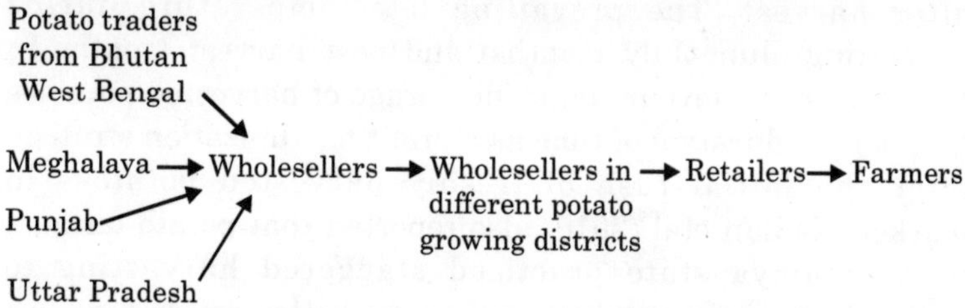

Fig. 5.1 : Frevailing marketing channel in Assam.

In East khasi hill district of Meghalaya, 'whole sellers in regulated markets' were reported to be utilized by more than 75 percent of the sampled farmers in the district, for disposing off the small, medium as well as mixed sized (unsorted) potato. In Meghalaya, farmers singly or in groups transport their potatoes to the nearby cities for marketing. However, for disposing the large sized potato more than 90 percent of the farmers utilized 'retailers in the traditional markets' (Table 5.6). In addition, 'assemblers coming to the villages' were the least utilized (5-10% of sampled farmers) in the district. The reason for this pattern of potato disposal could be the higher consumer preference thus higher market price for large size potato in the traditional markets, which was also the local vegetable market in Shillong city. The prevailing marketing channel has been depicted as in Fig. 5.1, Fig. 5.2 and Fig. 5.3.

Table 5.5 : Marketing channels utilized by farmers for disposal of table potato in Assam

Potato Size	Channels	Nagaon District			Barpeta District		
		No.	%	Rank	No.	%	Rank
Small	1. local traders in village	09	22.50	I	09	22.50	I
Medium	1. local traders in village	22	55.00	II	31	77.50	I
	2. Assemblers coming to village	27	67.50	I	31	77.50	I
	3. Wholesalers in regulated market	05	12.50	V	07	17.50	III
	4. Direct to city markets	07	17.50	IV	08	20.00	II
	5. Whole sellers coming to collect produce under contract	09	22.50	III	08	20.00	II
Large	1. local traders in village	22	55.00	II	28	75.00	II
	2. Assemblers coming to village	30	75.00	I	33	82.50	I
	3. Wholesalers in regulated market	07	17.50	IV	21	52.50	III
	4. Direct to city markets	09	22.50	III	10	25.00	IV
	5. Whole sellers coming to collect produce under contract	09	22.50	III	06	15.00	V
Mixed	1. local traders in village	27	67.50	I	31	77.55	I
	2. Assemblers coming to village	26	65.00	II	18	45.00	II

Table 5.6 : Marketing channels utilized by farmers for disposal of table potato in Meghalaya

Circles	*Size of potato sold*	*Channels*	*No of Farmers*	*Rank*
1	*2*	*3*	*4*	*5*
Mawphlang (n = 20)	Small	1. Whole seller in regulated markets	17(85.00)	I
		2. Retailers in traditional markets in city	01(05.00)	III
		3. Local village markets	10(50.00)	II
	Medium	1. Whole seller in regulated markets	18(90.00)	I
		2. Retailers in traditional markets in city	06(30.00)	II
		3. Local village markets	05(25.00)	III
	Large	1. Whole seller in regulated markets	06(30.00)	II
		2. Retailers in traditional markets in city	18(90.00)	I
		3. Local village markets	02(10.00)	III
	Mixed	1. Whole seller in regulated markets	16(80.00)	I
		2. Retailers in traditional markets in city	03(15.00)	III
		3. Local village markets	06(30.00)	II
Smit	Small	1. Whole seller in regulated markets	16(80.00)	I
		2. Retailers in traditional markets in city	03(15.00)	III
		3. Local village markets	14(70.00)	II
	Medium	1. Whole seller in regulated markets	15(75.00)	I
		2. Retailers in traditional markets in city	05(25.00)	III
		3. Local village markets	08(40.00)	II

(Contd...)

1	*2*	*3*	*4*	*5*
	Large	1. Whole seller in regulated markets	04(20.00)	II
		2. Retailers in traditional markets in city	19(95.00)	I
		3. Local village markets	00(00.00)	–
	Mixed	1. Whole seller in regulated markets	16(80.00)	I
		2. Retailers in traditional markets in city	00(00.00)	–
		3. Local village markets	05(25.00)	II

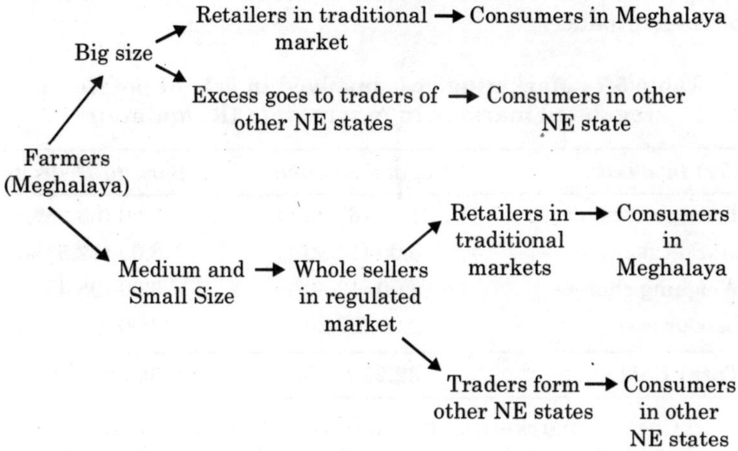

Fig. 5.2 : Marketing channels of potato produced in Meghalaya

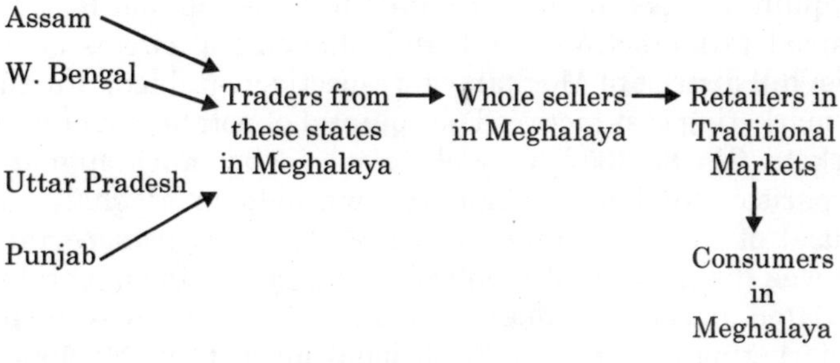

Fig. 5.3 : Marketing channels of potato Imported from other States in Meghalaya

Marketing Cost of Potato in Assam and Meghalaya

From the data presented in Table 5.7, it could be observed that the average marketing cost per quintal of potato for sampled farmers in Nagaon and Barpeta district of Assam was Rs 32.25 and Rs 35.50, respectively. Further, it could also be observed that about 66 percent (Rs 21.25/q) and 63 percent (Rs 22.50/q) of the total marketing cost was incurred on transportation of potato from villages to the regulated markets in Nagaon and Barpeta districts, respectively. In additions, market fee of Rs 5 and 8 per quintal was charged from farmers for selling per quintal of potato in the regulated markets. The high marketing cost involved in sale of potato could be the reason for preference among the sampled farmers for selling the produce to assembler or local trader in villages for disposal of their produce.

Table 5.7 : Marketing cost involved in sale of potato in regulated markets in Assam state (Rs /quintal)

Cost involved	*Nagaon Districts*	*Barpeta Districts*
Transportation cost	21.25 (65.89%)	22.50 (63.38%)
Market fee	5.00 (15.50%)	8.00 (22.53%)
Weighing charges	4.00 (12.40%)	3.00 (08.45%)
Labour cost	2.00 (06.20%)	2.00 (05.63%)
Total Cost	**32.25 (100%)**	**33.50 (100%)**

Cost of marketing per quintal of potato in regulated market in Meghalaya state was observed to be much higher in comparison to that in Assam (Table 5.8). The marketing cost per quintal of potato was observed to be Rs 36 and Rs 51 for sampled farmers from Smit and Mawphlang circles of east khasi hill district of Meghalaya, respectively (table8). Further, the marketing cost incurred per quintal of potato in traditional markets (Rs 85-90/q) was observed to be much higher in comparison to that in regulated markets in Meghalaya. A perusal of the same table also reflected that transportation cost was the major cost involved in the marketing of potato in regulated markets in Meghalaya state. However, with respect to marketing of potato in traditional markets in Meghalaya, charges for unloading from truck and carrying to market place

(Rs 20/q), weighing charges (Rs 15/q) and for carrying potato from weighing godowns of the traders(Rs15/q) were the other cost involved that substantially increased the marketing cost for the farmers.

Despite the high marketing cost incurred in selling potato in traditional markets, majority of the sampled farmers were utilizing traditional markets for disposing their medium to extra large sized potato. The reasons for this could be the fact that traditional markets in Meghalaya were the main vegetable markets in the city that fetched a higher price for larger sized potatoes and also the market margins for the farmers were higher due to involvement of only retailers and farmers in the marketing chain of potatoes.

Table 5.8 : Marketing cost involved in sale of potato in markets in Meghalaya (Rs /quintal)

Cost Involved	*Mawphlang Block*		*Smit Block*	
	Regulated Markets	*Traditional Markets*	*Regulated Markets*	*Traditional Markets*
Transportation cost	35(97.22%)	30(35.29%)	45(88.23%)	40(44.44%)
Unlaoding and carrying to market	–	20(23.53%)	05(09.80%)	15(16.67%)
Market fee	1(0.277%)	05(05.88%)	–	05 (05.55%)
Weighing charges	–	15 (17.65%)	01(01.96%)	15 (16.67%)
Labour cost	–	15 (17.65%)	–	15 (16.67%)
Total cost	**36(100%)**	**85(100%)**	**51(100%)**	**90(100%)**

Sources of Table Potato utilized by Potato traders of Assam and Meghalaya

Potato markets of West Bengal, traders from Guwahati and local villages were observed to be the sources of table potato utilized by all the 16 potato traders sampled from Assam state. Potato from Meghalaya was utilized by least number of traders in Barpeta (02) and Nagaon (03) in Assam (Table 5.9). On the other hand, all the sampled potato traders of Meghalaya utilized West Bengal Markets (Dubri, Alipur), Guwahati market as well as local potatoes from Shillong as the sources for meeting the potato demand in the state (Table 5.10). From Punjab, table potato was reported to be primarily imported in Meghalaya on commission basis (90% of the import), wherein potatoes were transported to the traders in Meghalaya, who after selling potatoes on prevailing market price and were returning the money to the farmers/traders in Punjab, after taking commission of six percent.

Table 5.9 : Sources of table potato utilized by traders in Assam (for whole year)

Source	*Nagaon(n = 08)*			*Barpeta(n = 08)*		
	Frequency	*Percentage*	*Rank*	*Frequency*	*Percentage*	*Rank*
West Bengal	08	100.00	I	08	100.00	I
Local villages	08	100.00	I	08	100.00	I
Traders from Guwahati	08	100.00	I	08	100.00	I
Potato from Meghalaya	03	37.50	II	02	25.00	II

Table 5.10 : Sources of table potato utilized by traders in Meghalaya

Sources	*Frequency*	*Percentage*	*Rank*
West Bengal (Dhubri & Alipur markets)	08	100.00	I
Guwahati market	08	100.00	1
Local potato from Shillong	08	100.00	I
Faruukhabad market, Uttar Pradesh	06	75.00	II
Punjab	05	62.50	III (Commission basis)

Marketing of potato in neighboring north eastern states from Meghalaya and Assam

It was revealed from the Table 5.11 that maximum quantity of potato from Nagaon district Assam was disposed to Nagaland, followed by Meghalaya and Arunachal Pradesh. While potato from Barpeta district of Assam was mainly exported to Meghalaya followed by Tripura and Arunachal Pradesh. The reason for maximum intake of potato in Meghalaya state could be the reason that in Assam, potato is grown as the winter crop harvested in the months of February March, which is a lean potato season in Meghalaya coinciding with the potato planting time in the state causing a much demand for fresh harvested potatoes from Assam.

Table 5.11 : Ranking of different states on the basis of quantity of potatoes disposed from Assam (n = 16)

Sl.No.	*State*	*Nagaon District*	*Barpeta District*
1.	Nagaland	I	V
2.	Meghalaya	II	I
3.	Tripura	–	II
4.	Arunachal Pradesh	III	III
5.	Bangladesh	–	IV

Potato was observed to be exported to all the north eastern states from Meghalaya except for Arunachal Pradesh (Table5.12). Traders expressed that maximum quantity of potatoes was exported to Manipur, followed by Assam and Tripura and least quantity was exported to Mizoram. Potato traders of Meghalaya also revealed that potato was exported directly to Bangladesh from certain pockets of east khasi district of the state and also it was exported via, Tripura as well as Assam states to Bangladesh.

Table 5.12 : Ranking of different states on the basis of quantity of potatoes disposed from Meghalaya (n = 08)

Sl.No.	*State*	*Rank*
1.	Manipur	I
2.	Assam	II
3.	Tripura	III
4.	Nagaland	IV
5.	Mizoram	V
6.	Bangladesh	VI

Major bottlenecks in potatao marketing in Assam and Meghalaya

Some of the major constraints in marketing of potatao as experienced by the farmers are presented in the following two tables which are self explanatory.

Table 5.13 : Major marketing related problems experienced by the potato farmers of Assam State

Sl.No.	*Major Problems*	*Rank Order of Problems*	
1. Related to Regulated Market		Barpeta (N = 40)	Nagaon (N = 40)
(*i*)	High share of middleman	II	III
(*ii*)	Tough bargain for good price	II	I
(*iii*)	Mal weighment by traders	VI	VI
(*iv*)	High transportation cost	III	IV
(*v*)	High market fee	VII	VII
(*vi*)	Distant location of regulated market	IV	IV
(*vii*)	Lack of adequate infrastructure like godowns, sheds etc to keep unsold produce	VI	V
(*viii*)	Poor road conditions leading to poor market accessibility	I	VIII
(*ix*)	Shortage of transportationfacilities	IV	V
(*x*)	Lower price during harvest	I	II
2. Related to Traditional Market in city			
(*i*)	Low price of produce in market	I	I
(*ii*)	High transportation cost	II	III
(*iii*)	Distant location of markets	V	VI
(*iv*)	Mal-weighment by traders	III	II
(*v*)	Monopoly of traders while buying potato	IV	II
(*vi*)	Shortage of transportation facilities	V	V
(*vii*)	Poor road conditions leading to poor market accessibility	I	IV

Conclusions

Table 5.14 : Major marketing related problems experienced by the potato farmers of East Khasi hills district of Meghalaya

Sl.No.	*Major Problems*	*Rank Order of Problems*	
1.	**Related to Regulated Market**	Mawphlang (N = 20)	Smit (N = 20)
(i)	High price fluctuations	IV	I
(ii)	Tough bargain for good price	III	II
(iii)	Mal weighment by traders	I	III
(iv)	High transportation cost	II	V
(v)	High market fee	VII	VI
(vi)	Distant location of regulated market	V	IV
(vii)	Lack of adequate infrastructure like godowns, sheds etc	VI	V
2.	**Related to Traditional Market**		
(i)	High cost involved in selling (Market fee, labour charge)	I	II
(ii)	Limited time for transaction	IV	I
(iii)	Untimely payment	II	II
(iv)	Mal-practices of traders especially in weighment	III	III
(v)	Monopoly of traders while buying potato	II	V
(vi)	High transportation cost	III	III
(vii)	Distant location of market	V	IV

Potato crop assumes immense importance in the cropping system, dietary habits of the people as the farm economy of North eastern region of India. The lack of proper storage facilities in the region compel the farmers to dispose off majority of the produce immediately after harvest, more so in places where potato crop is grown as winter crop. Distant location of markets and poor road conditions leading to a high transportation cost and thus higher marketing cost for sale of potato for the farmers in the region. Potato demands within the region were observed to be met from states within the region like Assam and Meghalaya as well as neighboring West Bengal state. The major potato growing states like Assam and Meghalaya in the region were observed to be the supplying potatoes to the entire north eastern states.

The potato marketing that is primarily in the hands of middlemen or assemblers in the region need to be organized with potato farmers as the main stakeholders. Focusing attention on creating more number of storage and marketing facilities, providing incentives for entrepreneurs for potato marketing would on one hand ensure a better profit margin for farmers, while on the other, would encourage farmers to achieve a higher potato production and productivity figures in the region.

PART - 2

EXPERIENCES OF INNOVATIVE RESEARCH INITIATIVES

CHAPTER 6

Production of Quality Planting Materials of Potato under Controlled and Field Condition

The inadequate availability of disease free high yielding quality seed potato is the major limiting factor for low potato productivity in NE region. Moreover, as the region is unfit for breeder seed potato production, search for alternative technology is inevitable in this direction. Hence, quality seed production through micro propagation techniques demands prominence.

Production of microplants by micropropagation of various potato varieties especially Kufri Giriraj and Kufri Jyoti for further production of minituber under the controlled condition in polyhouse may be a viable technological option. Also, microtubers of same cultivars may also be utilized for production of minituber under controlled conditions. The obtained minitubers may be further multiplied for the production of basic seed/quality planting materials. At the same time, *ex-vitro* experimentations are important to standardize the proper plant density for production of quality planting materials from the obtained minituber. This section of the book, therefore, highlights some of the research initiatives in the direction of quality seed production through micro-propagation techniques and te related experiments for technology generation.

Micro Propagation of Potato

Initiation of Cultures

Cultures for micro propagation can be initiated from meristem tips or sprout tips of pathogen-tested tubers or by *in-vitro* plantlets received through germplasm exchange. Alternately, culture may be initiated from auxiliary stem segments of pathogen-tested plants. In either of the above methods, to initiate culture, 2-3 explants are inoculated in a test tube (150 × 25 mm) containing about 10-15 ml solid propagation medium. The propagation medium contains basal MS salt and vitamins, GA_3 (0.1mg-^{l}), Calcium pentothenate (2mg-^{l}), NAA (0.01mg-^{l}), sucrose (30g-^{l}) and is solidified with 0.8% agar.

In-vitro Propagule Multiplication

Newly initiated culture is incubated at 25± 2^{0}C under 16 hours photoperiod of 150-w/mm^2 fluorescent lights. After 15-20 days in this condition apical/auxiliary buds of the explants grow into shoots and attain a length of 6 cm. These well-grown shoots now become ready for subculturing. In general, two methods are employed for *in-vitro* propagule multiplication.

In-vitro *Multiplication of Plantlets*

Multiplication of in-vitro plantlets through nodal cuttings has tremendous potential. After every 20-25 days, in-vitro plantlets can be subcultured on fresh solid propagation medium. At every subculturing, a single *in vitro* plantlet yields on an average three nodal cuttings. Since micro-propagation can be carried out round the year under controlled conditions, 18 sub-culturing can be done in a year. Therefore, theoretically 3^{18} (387 million) plant-lets could be obtained in one year from single *in-vitro* plant. Alternate method of mass micropropagation is liquid culture, when sufficient numbers of *in vitro* plantlets are obtained on solid medium in test tubes; these are cut into larger pieces (each with 1-2 nodes) and inoculated in liquid medium. Such 8-10 pieces are added in 250-ml conical flask having 20-ml liquid propagation medium. Composition of this medium is same as that of solid propagation medium except that it does not contain agar. Auxiliary buds from stem pieces grown into new plantlets and fills the flask within 15-20 days. Some laboratories use shaker speed 120-130 revolutions per minute to incubate liquid cultures. This ensures proper aeration and rapid growth of plantlets. The obtained plantlets, after hardening, can be directly transplanted in glass house/screen house and allowed to grow into normal size plants, producing normal tubers. Alternatively, these plantlets can also be utilized, after manipulating the media and environment, for microtuber production in laboratory condition.

Micro Tuber Production

Microtubers are miniature tuber produced under sterile tuber-inducing *in-vitro* condition. Therefore, they are considered to be absolutely free of viruses and other diseases. These small, dormant tubers are particularly convenient for handling, storage and distribution. Unlike micropropagated plantlets, they do not need the time consuming hardening period n a green house, and can also be adapted easily to large scale planting in the field.

For microtuber induction the unspent liquid propagation medium is dispersed out of the container, and 40 ml of fresh tuber induction medium is poured under aseptic conditions. The induction medium is based on MS basal supplemented with 44.38mm BA and 80 g^{-1} sucrose, and culture are incubated under complete darkness at 20°C. Depending on the genotype used, microtuber begin to appear 1-2 weeks after incubation. Usually they develop 60-75 days of incubation depending on the genotype. In general, 15-20 microtubers with an average weight of 100-150 mg can be harvested from each conical flask/ magenta box. In view of their delicate structure and high value, it is preferred to plant microtubers in greenhouses.

The micro propagation of potato was initiated since August, 2002 at Meghalaya. Four tubes of Kufri Jyoti and Kufri Giriraj (each containing 3 virus free micro plants) were procured from tissue culture unit of crop improvement, CPRI, Shimla, in the month of August, 2002. Multiplication of these micro plants was carried out following standard procedure recommended by CPRI, Shimla. This includes semi-solid media (Murashige and Skoog, 1962) and supplemented with sucrose, calcium pantothenate, GA_3 and Naphthalene acetic acid. It has been observed that the growth of Kufri Giriraj was comparatively faster in comparison to Kufri Jyoti, therefore the number microplant population in Kufri Giriraj was more than in Kufri Jyoti. By the end of 1st fortnight of Feb, 2006, the total available micro plant population was 14,900 (Kufri Jyoti-490 and Kufri Giriraj-7100) microplants. About 120 flask (with the assumption that each flask will produce approx. 10

microtuber) are under process for induction of microtubers. Out of the available stock, 2000 microtubers were also harvested, moreover, microplants of Kufri Giriraj was experimented for spacing requirement under polyhouse condition. Similarly, the multiplication of Kufri Jyoti is also carried out based on the number of microplants available.

Minituber Production

Minitubers are produced from in vitro plantlets or microtubers after they have been planted at high density (from 100-200 plants per m^2) in beds in poly- or screen houses. The plants remain small and consequently small tubers are produced. The Minitubers thus produced, can be planted directly in the field and enter the seed production chain. These Minitubers are classified as pre-basic seed. The advantage however, is the fact that the tubers obtained have been less exposed to infections with diseases than the clones that have already been in the field for a number of generations. This may be of particular importance in NE region where high degeneration rates prevail.

For quick generation of disease free seed material and to initiate the work on this line, 500 micro-tubers of each variety Kufri Giriraj and Kufri Jyoti was procured from tissue culture unit CPRI, Shimla in August 2002 and 2005. These micro tubers were subjected to multiplication under controlled conditions (poly house). It has been noted that the percent survival till harvesting was 74 % in case of Kufri Giriraj and 24% in case of Kufri Jyoti. Harvested minitubers of Kufri Giriraj and Kufri Jyoti were planted under polyhouse condition for their subsequent multiplication, and multiplication in field condition.

About 350 potato micro plants of Kufri Giriraj from tissue culture laboratory, CPRS, Shillong were also planted (after 10 days hardening) at spacing 10cm × 10cm, 15cm × 15cm and 20cm × 20cm under polyhouse condition to have preliminary information on optimum spacing for direct transplanting of microplant for minituber production. The status of quality seed produced from the laboratory since its inception and execution under this mission has been depicted in the following Table 6.1and 6.1a.

Table 6.1 : Status of micro plants/micro tubers/minitubers production (Field generation wise) Up to 15th February, 2006

Cultivars	*Micro Plants (Numbers)** (2006)*	*Micro Tubers (Numbers) (2006)*	*Minitubers (in kg) (2005-06)*	*Field Generation-I (in kg) (2004-05)*	*Field Generation II (in kg) (2003-05)*	*Field Generation III (in kg) (2003-05)*	*Field Generation IV (kg) (2003-05)*	*Total of all filed Generation (in kg)*
K.Giriraj	7100	4100	69.00	552.00	2614.00	454.00*	210.00*	3830.00
K.Jyoti	400	–	2.50	221.00	78.00	227.00	–	526.00
K.Kanchan	3200	–	12.80	17.00	–	–	–	17.00
K.Megha	100	–	–	–	–	–	–	–
K.Lauvkar	550	–	1.00	38.00	–	–	39.00	
K.Chandra–mukhi	350	–	1.40	32.00	–	–	–	32.00
K. Pukhraj	600	–	2.80	33.00	–	–	–	33.00
K. Ashoka	600	–	1.70	10.00	–	–	–	10.00
K. Badsah	250	–	0.60	1.00	–	–	–	1.00
K. Bahar	150	–	0.50	26.00	–	–	–	26.00
K. Sutlej	600	–	5.30	6.50	–	–	–	6.50
K. Chip–1	200	–	0.90	15.00	–	–	–	15.00
K. Lalima	300	–	2.80	–	–	–	–	–
K. Sindhuri	300	–	–	1.80	–	–	–	1.80
K. Chip–2	200	–	0.70	–	–	–	–	–
Total	14900***	4100	102.00	953.30	2692.00	681.00	210.00	4537.30

*600 Kg of Seed sold to Govt. of Sikkim (2006), **5000 microplants available in Laboratory. 2000 microplants under polyhouse.

Table 6.1a : Status of micro plants/micro tubers/mini–tubers production from April 2007 to Sept.2007

Cultivars	*Micro Plants (Numbers)*	*Micro Tubers (Numbers)*	*Mini-Tubers (in kg)*	*Field Generation I (in kg)*	*Field Generation II (in kg)*	*Field Generation III (in kg)*	*Field Generation IV (in kg)*	*Total of all field Generation (in kg)*	*Speed Supplied (Kg)*
K. Giriraj	6000	–	115.00	274.00	305.00	1375.00	11371.00	13325.00	3925.00
K. Jyoti	200	–	1.1	–	21.00	320.00	231.00	573.11	
K. Kanchan	4000	–	5.25	25.00	–	30.50	–		
K. Megha	150	–	–	–	–	–	–	–	–
K. Girdhari	50	–	–	–	–	–	–	–	–
K. Himalini	100	285	–	–	–	–	–	–	–
K. Lauvkar	200	–	1.50	3.5	12.00	34.50	–	51.50	–
K. Himsona	50	–	–	–	–	–	–	–	–
K. Pukhraj	200	–	1.50	6.00	22.00	32.50	–	62.00	–
K. Badsah	200	–	0.50	2.00	6.40	10.00	–	18.40	–
K. Bahar	150	–	1.5	4.00	11.00	13.00	–	28.00	–
K. Sutlej	200	–	3.30	10.50	9.00	14.50	–	34.00	–
K. Chipsona–1	150	–	3.50	4.00	15.00	13.25	–	32.25	–
K. Lalima	200	–	1.80	7.00	4.75	5.70	–	17.45	–
K. Sindhuri	150	–	1.50	2.80	1.60	5.00	–	9.40	–
K. Chipsona–2	100	–	2.50	6.60	3.50	2.50	–	12.60	–
Total	11850	1942*	138.95	345.40	411.25	1858.45	11602.00	14194.46	3925.00#

3925 Kg of Seed supplied to state department and for on–farm demonstrations (2007). 4700 micro plants available in Laboratory

* 1942 micro tubers procured from CPRI shimla and planted under polyhouse condition for minituber production.

On the other hand, another 900 micro plants were experimented to determine optimum N level and spacing under polyhouse condition. All the plant production and protection measures were followed as and when required after proper inspection. The experiment, therefore, was conducted under controlled condition with four level of nitrogen (0, 40, 80 and 120 Kg/ha.) and three spacing (20 × 15, 30 ×15 and 40 × 15 cm) each with three replication on potato microplants, which had been taken from tissue culture laboratory. The net plot size was 120 cm × 90 cm.

The mean survival % ranged from 76.39 to 81.37 and it was found non-significant. The plant height at 30 DAP not found significant in all the treatments and their interaction, where as at 60 and 90 DAP found significant with nitrogen level and spacing.

Table 6.2 : Effect of Nitrogen level and spacing on Survival % of Potato Microplants

Treatments	*S1*	*S2*	*S3*	*Mean*
N0	82.40	75.47	71.29	76.39
N50	78.70	82.88	76.85	79.48
N100	76.88	85.21	79.63	80.57
N150	77.77	81.94	79.63	79.78
Mean	78.94	81.37	76.85	

CD (P<0.05) N – NS S-NS N × S - NS

S1 = spacing 1, S2 = spacing 2 and S3 = spacing3

Application of 120 Kg N gave highest plant height (53.47cm) with 40 × 15 cm spacing at 60 DAP and 90 DAP also found maximum plant height with 120 Kg N and highest spacing. The weight of tuber per plant found significant with spacing. Highest mini tuber weight (233.40g) was found with S3 spacing and also highest weight (178.05g) with 150 Kg N. Average minituber weight was found to vary significantly with nitrogen, spacing and their interaction.

Table 6.3 : Response of nitrogen doses and spacing on plant height (cm) at 30, 60 and 90 Days after Planting (DAP)

Treatment	*30 DAP*				*60 DAP*				*90 DAP*			
	S1	S2	S3	Mean	S1	S2	S3	Mean	S1	S2	S3	Mean
N0	9.57	10.20	10.86	10.21	29.00	30.93	32.23	30.72	32.33	33.80	36.30	34.14
N40	11.80	12.10	11.20	11.70	28.90	36.47	42.60	45.66	30.56	38.63	43.40	37.53
N80	13.67	10.27	12.37	12.10	38.70	49.16	49.73	45.86	38.87	48.96	50.93	46.25
N120	12.40	11.77	10.96	11.71	46.80	50.40	53.47	50.22	46.80	50.56	53.96	50.44
Mean	11.86	11.08	11.35	35.85	41.74	44.51	37.14	42.99	46.15			
CD ($P<0.05$)	N- NS S- NS N × S–NS				N-3.63 S- 3.14 N × S –NS				N-3.37 S-2.92 N × S -NS			

Table 6.4 : Response of nitrogen doses and spacing on No. of tuber/plant

Treatments	*S1*	*S2*	*S3*	*Mean*
N0	4.43	4.85	7.08	5.45
N40	4.81	4.97	5.33	5.04
N80	5.21	5.85	7.35	6.14
N120	5.70	6.02	6.03	5.92
Mean	5.04	5.42	6.45	
CD (P<0.05)	N-0.59	S- 0.51	N × S – 1.02	

Table 6.5 : Response of nitrogen doses and spacing on wt. of tuber/plant

Treatments	*S1*	*S2*	*S3*	*Mean*
N0	112.58	148.39	217.00	159.32
N40	118.29	159.36	234.83	170.83
N80	125.00	129.06	252.01	168.69
N120	117.98	186.39	229.77	178.05
Mean	118.46	155.80	233.40	
CD (P-0.05)	N-NS	S- 35.63	N × S -NS	

Table 6.6 : Response of nitrogen doses and spacing on avg. tuber weight

Treatments	*S1*	*S2*	*S3*	*Mean*
N0	25.10	29.43	28.55	27.69
N40	24.11	31.46	39.28	31.62
N80	30.45	31.35	32.82	31.54
N120	26.69	33.58	42.24	34.17
Mean	26.59	31.45	35.72	
CD (P-0.05)	N-2.53	S-2.19	N × S -4.38	

The minituber weight (g) was 26.59, 31.45 and 35.72 with S1, S2 and S3 spacing, respectively. Highest level of N gave maximum (34.17g) avg. minituber weight.

Minituber yield was found to be significantly varying with doses of nitrogen, spacing but their interaction was non-significant. The maximum minituber yield (3.54 kg/plot) was found with spacing 20 × 15 cm followed by 3.38 kg and 3.30 kg per plot with S2 and S3, respectively.

Moreover, the differential experiences of quality seed production through microtuber or microplants were also experienced which have been depicted in Table 6.8

Table 6.7 : Response of nitrogen doses and spacing on tuber yield (kg/plot)

Treatments	*S1*	*S2*	*S3*	*Mean*
N0	3.18	2.75	2.55	2.83
N40	3.10	3.17	3.03	3.10
N80	4.07	3.90	3.77	3.91
N120	3.80	3.68	3.84	3.77
Mean	3.54	3.38	3.30	
CD (P-0.05)	N-0.27	S – NS	N × S-NS	

Table 4.8 : Comparative performance of microtubers and microplants for minitubers production under polyhouse condition

Parameters	*Microtubers*	*Microplants*
Time required for production and planting	7-8 months	One month
Cost involved (growing media, induction hormones, space and manpower required)	More cost	Less cost
Transportation	Easy to carry	Not suitable
Dormancy	3 months	No dormancy, ready for direct transplanting at any time
Growth and Yield parameters		
Plant height (at 90 DAT)	90.53 cm	59.40cm
Number of primary branches/plant	7.53	3.54
Yield per plant	178.0g	198.0g
Number of tubers per plant	17.0	13.86
Average tuber weight	10.58g	16.98g

CHAPTER 7

Field Experiments on Minituber and Other Sources of Seed Potatoes under Rainfed Conditions of Meghalaya

Several field experiments pertaining to minitubers, microplants, etc were conducted in comparison to TPS tuberlets, as well as other potato cultivars like Kufri Jyoti and Kufri Giriraj. Similarly, effect of poultry manure was also ascertained on performance of microplants under controlled condition. Optimum plant density for microplants was also standardized for maximum production of minituber under polyhouse condition. The details of theses experiments and findings thereupon are presented in the following sub heads :

Performance of mini-tuber and TPS Tuberlets compared to Kufri Jyoti and Kufri Giriraj

Table 7.1 : Percent emergence at 45 Days

Population/Var.	*RI*	*RII*	*RIII*	*RIV*	*Total*	*Average*
MT. Gen. I*	72.2	81.9	81.9	83.4	319.4	79.8
HPS-7/MST-1	66.6	66.6	63.8	62.5	259.5	64.9
TPS C-3	83.4	73.6	76.4	83.4	316.8	79.2
HPS-1/13	80.5	72.2	70.8	77.8	301.3	75.3
HPS-7/67	72.2	72.2	58.3	69.4	272.1	68.0
Kufri Giriraj	54.1	52.7	52.7	55.5	215.0	53.7
Kufri Jyoti	91.6	72.2	66.6	76.4	306.8	76.7

The performance of the produce of first generation microtubers (minitubers) was tested in comparison to the TPS tuberlets and standard cultivar Kufri Jyoti and Kufri Giriraj. Findings of the experiments revealed that the yield of the minituber was considerably ahead (213.9 q/ha) of the yield obtained from Kufri Jyoti (136 q/ha) and Kufri Giriraj (168.32 q/ha) as indicated in the Table 7.2. Further, the yield performance of TPS C3 (202.75 q/ha) was comparable to the yield obtained from first generation minitubers. The percentage emergence also varied accordingly (Table 7.1). However, the dry matter contents was highest in the TPS hybrid HPS 7/67 followed by TPS C3 and HPS 7/MST-1 (Table 7.3).

Table 7.2 : Tuber Yield (q per hectare)

Population/Var.	*RI*	*RII*	*RIII*	*RIV*	*Total*	*Average*
MT. Gen. I*	218.75	200.23	231.48	205.21	855.67	213.92
HPS-7/MST-1	115.64	123.84	130.89	123.38	493.75	123.44
TPS C-3	187.08	221.48	208.33	194.10	810.99	202.75
HPS-1/13	152.73	166.67	155.14	158.22	632.76	158.19
HPS-7/67	174.61	167.99	155.10	165.86	663.56	165.89
Kufri Giriraj	172.87	166.67	165.51	168.22	673.27	168.32
Kufri Jyoti	145.25	132.52	130.21	136.00	543.98	136.00

Table 7.3 : Percent Tuber Dry Matter

Population/Var.	*RI*	*RII*	*RIII*	*RIV*	*Total*	*Average*
MT. Gen. I*	16.84	16.00	15.73	15.00	63.57	15.89
HPS-7/MST-1	17.60	17.25	18.46	17.30	70.61	17.65
TPS C-3	18.91	16.63	18.10	18.00	71.64	17.91
HPS-1/13	16.45	16.58	17.05	16.80	66.88	16.72
HPS-7/67	19.45	17.89	19.88	18.30	75.52	18.88
Kufri Giriraj	15.60	14.43	16.45	14.50	60.98	15.24
Kufri Jyoti	16.80	16.50	16.02	17.00	66.32	16.58

*Produce of first generation micro tuber (minituber)

Effect of poultry manure with different combination of inorganic fertilizers for quality seed potato production by utilizing minitubers (cv. Kufri Giriraj) under rainfed conditions of Meghalaya

The on-station experiment was conducted during 2003 and 2004. The soil of the experimental site was sandy clay loam in texture and acidic having PH 5.2,N 202 kg/ha, P 58 kg/ha and K 185 kg/ha. The experiment was laid out in Randomized Block Design with four replications. The net plot size was 7.2 sqm. There were six treatments *viz.* 100 % through inorganic fertilizers(T1), 25 % poultry manure +75% through inorganic (T2), 50% poultry manure +50% through inorganic(T3), 75% poultry manure +25% through inorganic(T4), 100% through poultry manure(T5), 125% through poultry manure (T6). All the treatments were applied with uniform doses of 120 kgN (in two split doses), 120 kg P_2O_5 and 60 kg K_2O/ha. The nitrogen, phosphorus and potash were applied through urea, single super phosphate (SSP) and muriate of potash (MOP), respectively. The crop was raised as per recommended package of practices for seed multiplication in the region. The plants were dehaulmed at 105 days after planting (DAP). The plants were harvested after 15 days of haulm cutting and data were recorded grade wise (Large, medium and small). All the parameters like germination percentage, plant height, number of branch/hill and yield were statistically analyzed according to the standard methods (Gomez and Gomez, 1984).

Table 7.4 : Effect of poultry manure on growth and yield of potato

Treat-ments	*Percentage Germination*		*No of Hill / Plant*		*Plant Height (cm)*		*Tuber Yield (q/ha.)*	
	2003	*2004*	*2003*	*2004*	*2003*	*2004*	*2003*	*2004*
T1	98.74	97.50	6.70	5.95	49.25	49.60	232.96	240.71
T2	98.75	96.60	6.47	6.23	51.20	51.90	226.87	238.67
T3	97.50	96.60	5.27	5.46	49.00	51.00	207.90	213.67
T4	98.14	95.40	5.65	5.75	50.52	49.97	188.66	193.14
T5	97.10	96.70	5.87	5.87	48.21	47.68	171.39	181.04
T6	98.10	94.57	5.40	5.48	49.00	49.48	178.48	190.32

CD 0.05 Y=NS T=NSY=NS T=NSY=NS T=NS Y=NS T=12.06

Findings revealed that the percentage plant emergence was appreciably high and ranged from 94.57 to 98.75. The germination percentage were good (Table. 7.4). Germination percentage was not influenced by any of the treatments, year and their interaction. Similarly, plant height and number of branches/hill were also not found influenced by any of the treatments.

Table 7.5 : Effect of poultry manure on yield (q/ha) of different grade of tubers

Treat	*2003*			*2004*		
	Grade A	*Grade B*	*Grade C*	*Grade A*	*Grade B*	*Grade C*
T1	117.36	95.83	19.44	125.42	98.89	16.40
	(50.37)	(41.11)	(8.52)	(52.10)	(41.00)	(6.90)
T2	110.30	101.75	14.81	106.26	107.65	14.75
	(48.62)	(44.85)	(6.53)	(44.52)	(45.10)	(10.38)
T3	106.25	94.50	17.14	92.89	98.69	16.70
	(51.11)	(45.45)	(3.44)	(43.47)	(46.19)	(9.27)
T4	80.30	96.10	12.24	83.25	97.20	17.25
	(42.51)	(50.93)	(6.50)	(43.10)	(50.33)	(6.57)
T5	64.28	99.22	13.88	60.66	100.06	19.82
	(37.50)	(57.89)	(4.61)	(33.51)	(55.26)	(11.23)
T6	62.25	102.36	14.81	63.61	104.76	21.75
	(34.87)	(57.35)	(7.78)	(33.42)	(55.04)	(13.54)

Figures in parenthesis indicate the percent yield of respective grade over the total yield

Parameters	*Grade A*	*Grade B*	*Grade C*
Year(Y)	NS	NS	NS
Treatment (T)	12.67	NS	NS
Y × T	**NS**	**NS**	**5.02**

Results further indicated that the various treatment combinations had significant difference in tuber yield. It also showed that there was no significant difference over year and their interaction with different treatments. However, the difference in yield of T1 and T2 was found to be statistically at par. The lowest yield (171.39 and 181.04 q/ha.) was recorded in the treatment T5 (100% PM) in both the years. Increasing rate of PM from 100% to 125 % gave higher yield but remained statistically at par to each other. The similar results were also reported by Uppadhay *et al.* 2003, where they had substituted inorganic fertilizer with vermicompost in potato production. From the result (Table 7.5) on grade wise tuber yield, it was found that there was significant difference in yield of grade A (>50 g) tubers among different treatment combinations. The highest tuber yield of grade A was recorded 117.36 and 125.42 q/ha in the treatment (T1) in the year 2003 and 2004, respectively. However, the lowest yield 62.25 q/ha was recorded in (T6) in the year 2003 and during 2004 the lowest yield 60.66 q/ha was recorded in the treatment T5. Similarly, the yield of grade B (30-50 g) tuber was not affected significantly by any of the treatments. The highest yield of grade B tuber during 2003 and 2004 was recorded 101.75q/ha and 107.65q/ha, respectively in the treatment T2. The yield of grade C (<30 g) tuber was also found to be statistically non-significant across the treatments and over the year, but their interaction (Y × T) showed significant difference in yield. Table 7.5).

It was also noted from the same table that there was increasing trend of percent medium size tubers with increasing dose of poultry manure in combination with inorganic fertilizers and 100% poultry manure alone. However, this trend was slightly decreased when 125% poultry manure was applied. Finding of study helped to conclude objectively that the inorganic fertilizers could be substituted with the organic sources like poultry manure to the extent of 25 % for better sustainable management of soils of NEH region without any significant yield loss over inorganic fertilizers.

Effect of plant density on minituber production from micropropagated plants

The use of *in vitro* plantlets for production of pre basic seed potatoes was evaluated in order to assess the appropriate planting density under polyhouse condition. *In vitro* plantlets of cv. Kufri Giriraj were planted in three-plant densities *viz.*, 25, 36 and 81 m^{-2}. Percent survival across all the densities was appreciably high (93.5-99.6). The number of compound leaves, tubers per plant and average tuber weight varied significantly in different plant densities. (Table 7.6)

Table 7.6 : Effect of plant densities on yield and yield related attributes of microplants

Plant Density m^{-2}	% Sur-vival	Plant Height (cm)	No. of Compd. Leat/ Plant	No. of Prim. Bran-ches $Plant^{-1}$	No. of Tubers		Yiled		
					m^{-2}	$Plant^{-1}$	Kg m^{-2}	g $Plant^{-1}$	Ave. tuber Weight (g)
D1-25	94.7	36.0	7.0	4.0	140	5.9	2.80	118.48	19.98
D2-36	93.5	35.0	5.0	3.6	179	5.3	2.85	87.75	15.93
D3-81	99.6	39.7	4.0	3.6	294	3.6	3.13	38.88	10.66
LSD 0.05	NS	1.51	1.31	NS	11.18	0.46	NS	20.10	3.41

The maximum tuber yield of 3.13 kg m^{-2} with an average tuber weight of 10.66 g was recorded in D3 followed by 2.85 kg with an average tuber weight of 15.93g in D2 and 2.80 Kg with an average tuber weight of 19.98 g in D1. However, yield across all the densities were statistically at par. It is evident from the above findings that the yields obtained in different plant densities were statistically at par; however, number of minitubers $plant^{-1}$ varied significantly. Plant density D1 gave highest number of tubers/ $plant^{-1}$ (5.9) with an average tuber weight (19.98 g) as compared to other two densities. (Table 7.6)

The above results categorically indicated that per unit minituber production efficiency was the maximum when the micropropagated plants were planted at a density of 36 plants m^{-2} (i.e. D2). Therefore it is suggested that the production of minitubers by following 36 plants m^{-2} would be more efficient way for rapid multiplication of pre-basic seed potato tubers during favourable weather under polyhouse condition at Shillong.

CHAPTER 8

Availability and Supply of Quality Seed of High Yielding Potato Varieties in Different North Eastern States

As potato is the main crop of this region and the non-availability of quality seed potato is the major limiting factor for low potato productivity in this region, it is, hence, apt to estimate the actual requirement of the same in this region (Table 8.1). Therefore, under this Mission, this station strive for supply of quality seed to various states both free of cost (Table 8.2) as well as enhanced supply from CPRS, Kufri (Table 8.3).

Table 8.1 : Requirement of Potato Seed ('000 t) in NE Region

State	*Av. Area ('000 ha)*	*Av. BS* Unit Required***	*AV. CS* Unit Required****
Arunachal Pradesh	4.30 (VI)	0.03	03.8
Assam	77.83 (I)	0.53	66.8
Manipur	4.83 (V)	0.02	02.5
Meghalaya	18.20 (II)	0.12	15.2
Mizoram	0.43 (VIII)	0.005	00.6
Nagaland	4.00 (VII)	0.03	04.2
Sikkim	6.80 (III)	0.04	05.2
Tripura	5.50 (IV)	0.04	04.5
Total	118.92	0.815	102.8

*BS = Breeder Seed; CS = Certified Seed.

Figures in parentheses indicate ranks

**BS to be multiplied 3 times (125 fold multiplication) to produce CS.

***Seed rate 2.5 t/ha; Seed replacement after every three years.

Table 8.2 : State–wise seed supply under various projects

State	*Varieties*	*Seed supplied year-wise (in q)*					*Total*
		2002–03	*2003–04*	*2004–05*	*2005–06*	*2006–07*	*(in q)*
Arunachal Pradesh	K. Jyoti	10.0	–	–	30.0	40.0	80.0
K. Giriraj	10.0	–	–	–	–	10.0	
Assam	K. Jyoti	–	08.0	–	–	–	08.0
K. Giriraj	–	–	40.0	–	23.0	63.0	
Manipur	K. Jyoti	–	40.0	–	–	–	40.0
K. Girirai	–	–	10.0	-	–	10.0	
Meghalaya	K. Jyoti	100.0	92.0	–	10.0	28.0	230.0
K. Giriraj	25.0	–	–	–	–	25.0	
K. Megha /	16.0	16.0					
Mizoram	K. Jyoti	10.0	–	–	–	–	10.0
K. Girirai	20.0	–	20.0	–	–	40.0	
Nagaland	K. Jyoti	25.0	–	–	15.0	–	40.0
K. Giriraj	05.0	–	20.0	–	–	25.0	
Sikkim	K. Kanchan/	–	08.0	–	–	3.0	11.0
K. Jyoti	–	–	–	17.0	–	17.0	
K. Giriraj	04.0	04.0					
Tripura	K. Jyoti	–	06.0	–	–	–	06.0
K. Giriraj	–	–	10.0	–	–	10.0	
TOTAL	205.0	154.0	100.0	70.0	110.0	645.0	

(–)No supply

The above quantity of seed if multiplied thrice shall cover about 28,667 hectares of area (about 20% of the total area under potato) under quality seed potato in the region.

Besides, AAU, Jorhat has procured quality seed of Kufri Jyoti (10q), Kufri Megha (30 qt); Assam Seed Corporation (ASC) has taken 40 qtl of Kufri Jyoti. In addition, ICAR Research Complex for NEH Region, Umiam and Govt. of Meghalya have procured 8qtls and 25 qtls of quality seed of cultivar Kufri Jyoti, Kufri Giriraj and Kufri Megha for use in their respective watershed areas.

Table 8.3 : Year-wise seed allotment and supply to NEH States

Year	*Allotment (q)*	*Supply (q)*
2000- 01	373.00	180*
2001- 02	425.00	265*
2002- 03	370.00	319*
2003- 04	435.00	350*
2004- 05	452.00	436.22**
2005-06	433.00	310*
2006-07	445.00	375*
Total	**2933.00**	**2235.22**

* Did not lift the allotted quantity.

** Short supply due to low production.

PART - 3

Technological Backstopping for Potato Production

CHAPTER 9

Suitable TPS Hybrids/Lines Hybrids for NE Region

True Potato Seed (TPS) as the low cost alternative to seed tuber has been established by using TPS derived planting material (CIP, 1991; Chadha, 1994). Most widely assessed TPS derived planting material are F_1 seedlings and seedling tubers. Analysis of TPS adaptability trials has revealed that TPS progenies compete favorably with prevailing clonal checks (Kadian and Upadhya, 1994). The state of Meghalaya has tremendous potential for utilization of TPS as it has significant proportion (5.6%) of total cultivable land under potato. Moreover, lower productivity of potato (about 9 t/ha) coupled with remarkably high cost of production further offer the opportunity for introduction and dissemination of TPS technology in the state. Study was therefore, undertaken to assess TPS derived materials namely TPS seedlings of different hybrids and seedling tubers obtained thereupon under the rain fed condition of Meghalaya on the yield and the related attributes.

Nine TPS lines and hybrids were evaluated for seedling tuber production in summer season. Seedlings were raised both in the open field condition as well as the glass house to see the differential germination, if any. Data on the aspects of plant height and yield per plot and per hectare were recorded. Moreover, the obtained seedling tubers of two hybrids (92-PT-27 and TPS C-3) were planted in the subsequent season and yield data were recorded. The findings are contained in Table 9.1.

Table 9.1 : Evaluation of different TPS Hybrids/crosses for seedling tuber production.

S. No.	*TPS Hybrids/ lines*	*Germination %*	*Plant height at 75DAP*	*Yield kg/ plot*	*Yield (q/ha)*
1.	92-PT-27	75.00	60.50	6.48	60.0
2.	TPS C-3	72.00	57.8	6.10	56.2
3.	CP-3192XPRT-19	67.00	5504	5.60	52.0
4.	PRT-12XPRT-19	50.20	48.9	202	20.0
5.	PRT-2XPRT-6	48.50	47.8	1.62	15.0
6.	PRT-33XPRT-6	52.00	46.5	2.20	20.0
7.	PRT-20XPRT-19	45.60	47.0	1.62	15.0
8.	PRT-4XPRT-6	43.80	45.6	1.62	15.0
9.	PRT-3XPRT-6	57.50	50.5	3.78	35.0
	CD(0.05)	**4.20**	**3.24**	**0.70**	**13.8**

Out of the 9 TPS hybrids studied, the hybrid 92-PT-27 showed higher germination (75%) over the others (which ranged from 43.8-72.0%). However, in glasshouse condition 92-PT-27 showed 85% germination. Similarly, yield wise also, 92-PT-27 (60 q/ha) out performed the other hybrids/lines. In the next generation, seedling tubers of TPS hybrid 92-PT-27 (125q/ha) outperformed TPS C-3 (115q/ha) but lesser than Kufri Giriraj (252 q/ha).

Similarly, in another trial, four TPS hybrids were evaluated for seedling tuber production and commercial crop production through tuberlets in subsequent generation under the above programme. Experiment was laid out in Randomized Block Design (RBD) with six replications for assessing seedling crop. The TPS of four progenies (HPS 7/MST-7, HPS 1/13, HPS 7/67 and TPS C-3 [as the control]) were sown in nursery beds and the transplanting of seedling was done at 40 cm × 10 cm spacing after six weeks of sowing. Commercial crop through seedling tubers was raised following RBD with four replications. Well-sprouted seedlings tubers of more than 30 g size were planted at the spacing of 60cm × 20cm. The same size of tubers of K. Jyoti and K. Giriraj were used as control. Chemical fertilizers, well rotten FYM and all the cultural practices recommended for this region at various stages of growth were followed. However, only one spray of fungicide (Mancozeb) was given to the crop during both the year.

Table 9.2 : Performance of Different Progenies as F1 Seedlings

TPS Hybrids	*Survival Percentage*	*Total Yield (q/ha)*	*Marketable Yield (q/ha)*	*Uniformity*
HPS-7/MST-1	83.76	47.08	19.02 (40.40)	3.0
TPS C-3	84.04	38.87	18.74 (48.21)	4.0
HPS-1/13	81.88	35.96	16.39 (45.58)	3.5
HPS-7/67	88.99	64.59	30.55 (47.30)	3.0
SEm±	2.23	7.41	3.62	–
CD (95 %)	4.76	15.80	7.71	–
CD (99%)	6.59	21.86	10.68	–

Figures in parenthesis indicate percentage of total yield.

Table 9.3 : Performance of Seedling Tuber

TPS Hybrids Cultivars	*Percent Emergence*	*Total Yield (q/ha)*	*Percent Tuber dry Matter*	*Incidence of Late Blight (%)*
HPS-7/MST-1	69.60	122.37	17.65	25
TPS C-3	82.57	203.60	17.91	25
HPS-1/13	77.07	156.07	16.72	50
HPS-7/67	70.52	164.95	18.88	50
K. Giriraj	69.27	164.05	15.24	50
K. Jyoti	78.75	136.60	16.58	90
SEm±	2.77	5.06	0.42	–
CD (95 %)	5.90	10.78	0.89	–
CD (99%)	8.17	14.93	1.24	–

Similarly, total yield of HPS 7/67 (64.59 q/ha) was found significantly out performing HPS 7/ MST-1 (47.08 q/ha), TPS C-3 (38.8 q/ha) and HPS 1/13 (35.96 q/ha) and accordingly marketable yield varied in that order i.e. HPS 7/67 (30.55 q/ ha) and HPS 1/13 (16.39 q/ha) (Table 9.2). It could be noticed from the same table that the percent of marketable yield was marginally better in TPS C-3 (48.25%) than HPS 7/67 (47.30%), HPS 1/13 (45.58%) and HPS 7/ MST-1 (40.40%). The lower yield from TPS progenies could be attributed to the dependence of TPS crop on rainwater only as no irrigation was supplied over and above rainfall.

Findings of above studies helped to conclude that 92-PT-27 outperformed HPS 7/67 and emerged as best TPS progeny for utilization as F1 seedling crop in the rainfed condition of Meghalaya. However, HPS 7/67 was also found comparable with TPS C-3 both as F1 seedlings and seedling tuber planting material. Moreover, yield from seedling tuber was found competing well with popular potato cultivars namely K. Jyoti and K. Giriraj. Findings also helped to derive that there was cost saving from reduced fungicide use. Hence, the yield and disease advantages from TPS derived planting materials would also induce the potato farmers to use F1C2 and F1C3 seed tubers for subsequent crops consequently reverting from the traditional tuber seed system.

Based on the above identification of suitable TPS hybrids like 92-PT-27, HPS 7/67 and TPS C3, further on-farm and on-station trials were conducted in both summer and autumn season under the rainfed condition of Meghalaya. The details of this experiments as well as consequent findings are presented in the subhead 6.4 of programme 3 of the document at hand.

As a result, TPS hybrids namely, 92-PT-27 and TPS C 3 were finally utilized for the large scale dissemination among the potato growers of NE states through on-farm demonstration, etc.

CHAPTER 10

Evaluation of Red Skinned Potato Cultivars/Hybrids

Red skinned potatoes have higher marketable value, much in demand for consumption in north central and eastern plains of India. Besides, they are preferred in neighboring countries like Bangladesh, Bhutan, Nepal, Pakistan and Philippines (Pandey *et al.*, 2000). In this region also red skinned cultivars are preferred over white skinned cultivars especially in the states of Sikkim, Darjeeling (northern hills of West Bengal) and part of Manipur. Keeping the above facts in view and considering demand of red skinned potatoes in this region two advanced red-skin potato hybrids and three red-skin commercial varieties including Desiree were evaluated for their potential use in NEH region using cultivers Kufri Giriraj and Kufri Jyoti as control.

The on-station experiment was laid out with five red skinned genotypes in a randomized block design with four replications. The plot size of the experiment was 7.2 m^2, with row to row and plant to plant spacing of 60cm × 20cm. Recommended package and practices were followed to raise the crop under rainfed conditions. All the genotypes were given uniform doses of 120 kg N (two splits), 120 kg P_2O_5 and 60 kg K_2O/ha and all the plant protection measures were undertaken except for late blight. Late blight assessment was done at weekly intervals as per British Mycological Society (BMS) scale (Anonymous, 1947). Area under disease progress curve (AUDPC) was calculated (Abbot, 1973) to compare different treatments. The crop was harvested at maturity at 120 days after planting.

Table 10.1 : Growth parameters and incidence of late blight of red skinned cultivars as compared to Kufri Giriraj and Kufri Jyoti

Cultivars /hybrids	*Germination(%)*			*Plant Height (cm)*			*AUDPC*		
	2003	2004	Mean	2003	2004	Mean	2003	2004	Mean
K. Giriraj	90.00	91.67	90.83	65.93	65.50	65.72	177.04	244.71	210.87
K. Jyoti	91.25	96.66	93.95	57.55	58.95	58.25	564.38	510.42	537.40
MP/92-56	97.09	99.50	98.29	78.80	77.76	78.28	327.54	303.34	315.44
K. Lalima	98.34	98.72	98.53	66.20	64.77	65.49	392.30	378.88	385.59
Desiree	65.58	97.07	81.33	63.75	65.27	64.51	499.63	478.05	488.84
MS/92-2105	92.91	98.32	95.61	68.37	66.39	67.38	476.87	422.63	449.75
K. Kanchan	92.06	87.91	89.99	77.20	75.05	76.12	259.29	217.88	238.58

CD (P = 0.05) Y-5.25 V-6.75 YxV-9.55 Y-NS V-6.25 Y-11.88 V-22.23 YxV-NS

AUDPC- Area under disease progress curve

Table 10.2 : Yield of red skinned cultivars against Kufri Giriraj and Kufri Jyoti

Cultivars/hybrids	*Total yield (q/ha)*			*Marketable yield (q/ha)*		
	2003	*2004*	*Mean*	*2003*	*2004*	*Mean*
K. Giriraj	226.05	201.77	213.91	196.45 (86.90)	177.75 (88.90)	187.10 (87.47)
K. Jyoti	121.55	208.87	165.21	107.21 (88.20)	170.53 (81.64)	138.87 (84.06)
MP/92-56	65.73	74.32	70.03	55.38 (84.25)	52.73 (70.94)	54.06 (77.19)
K. Lalima	117.81	153.30	135.55	101.74 (86.35)	127.19 (83.43)	114.46 (84.44)
Desiree	99.68	166.12	132.90	87.85 (88.13)	148.33)(88.29)	118.10 (88.86)
MS/92-2105	67.03	85.62	76.32	53.17 (79.32)	71.33 (88.30)	62.25 (81.56)
K. Kanchan	66.90	87.20	77.05	53.23 (77.98)	58.43 (75.83)	55.83 (71.79)

CD (P = 0.05) Y-27.45 V-23.88 YxV-33.34 Y-NS V-22.17 YxV-31.35

Figures in the parenthesis are percentage of marketable tubers over total yield

Data in the Table 10.1 indicate that percentage germination at 45 DAP ranged from 65.58 to 98.34 per cent and 87.91 to 99.50 per cent in the year 2003 and 2004 respectively. The maximum and significant mean germination percentage was recorded in Kufri Lalima (98.53) and significant minimum percentage (81.33) was recorded in Desiree. The germination percentage was also affected significantly by year and their interaction with genotypes (Y × V).

There was significant difference on mean plant height among the different genotypes. The maximum and significant average plant height at 75 DAP was recorded in MP/92-56 (78.28 cm) followed by Kufri Kanchan (76.12 cm) and the minimum (58.25 cm) in Kufri Jyoti. The effect of year and interaction between genotype and year on mean plant height was found to be non-significant.

Data on AUDPC indicated that there was significant difference among the genotypes and within the years (Table 10.1). The least (210.87) and highest (537.40) mean AUDPC values were recorded in Kufri Giriraj and Kufri Jyoti respectively, which are white skinned varieties used as check. Among the red skinned cultivars, Kufri Kanchan performed the best, showing least mean AUDPC values (238.58) and was followed by MP/92-56 (315.44) and Kufri Lalima (385.59). However, the yields of the red skinned varieties were not positively correlated with their blighting rate (AUDPC). This may be due to the fact that some other factors *viz.*, the variable bulking rate of different genotypes might have played role in the above trend of results.

Table 10.2 shows that the mean marketable tuber yield ranged from 54.06 (MP/92-56) to 187.10 q/ha (Kufri Giriraj). Kufri Kanchan (55.83 q/ha) also recorded second lowest marketable yield. The maximum marketable tuber yield recorded in Kufri Giriraj, one of the white skinned late blight resistant cultivars used as check. However, mean marketable tuber yield among red skinned genotypes/advanced hybrids varied from 54.06 q/ha to 118.10 q/ha.

The significant and maximum mean marketable yield among red skinned genotypes was recorded in Desiree (118.10 q/ha) followed by Kufri Lalima (114.46 q/ha.). Percent mean marketable yield was maximum in cv. Desiree (88.86%) followed by Kufri Giriraj (87.47%) and Kufri Lalima (84.44%). The mean marketable yield was also affected by year and variety interaction (Y × V).

With regard to mean total tuber yield (Table 10.2) over both the years indicate that white skinned check cultivars Kufri Giriraj (213.91 q/ha) and Kufri Jyoti (165.21 q/ha) emerged significantly best. Among the red skinned cultivars / hybrids Kufri Lalima (135.55 q/ha) found to be significantly superior followed by Desiree (132.90 q/ha) and Kufri Kanchan (77.05 q/ha). However, there was significant difference in the total tuber yield obtained from different genotypes during both the years and their interactions. During the year 2003, the general performance of different genotypes might be affected due to more congenial environment for the spread and progress of late blight infection in comparison to the year 2004. However, in case of Kufri Giriraj the severity of late blight during 2004 was more in comparison to 2003, resulted in higher yield during 2003.

CHAPTER 11

Evaluation of Organic and Inorganic Sources of Nutrients on Potato Cultivars and Seedling Tubers

Modern Agriculture undoubtedly has increased production and labour efficiency. But, concerns have been raised from time-to-time against its adverse effect on soil productivity and environment. Dependence of Modern Agriculture on use of fossil based inputs, such as chemical fertilizers and pesticides/ herbicides as well as farm machinery are held impossible for the adverse effect.

Biofertilizers and other organic sources of plant nutrient are actually the preparations containing live or latent cells of efficient strains of beneficial microorganisms applied either to soil or as seed treatment. These organic based nutrients increases microbial count and accelerate certain microbial processes, augmenting availability of nutrients to growing plants. They include symbiotic nitrogen fixing Rhizobium, free-living nitrogen fixing Azotobacter and Azospirillum, Azolla, Blue Green Algae (BGA), Phosphorus Solubilising Bacteria (PSB) and phosphorus uptake enhancing Mycorrhiza. In INM, biofertilizers are used only to supplement nutrients in combination with chemical fertilizers and organic manures. Therefore, experiments were conducted to ascertain the relative efficacy of organic and inorganic source of nutrients on the yield and yield related attributes of different potato cultivars as well as TPS tuberlets.

Preliminary studies on effect of inorganic and organic sources of nutrients on the yield of potato cultivars

A field experiment was conducted to study the response

of potato varieties to organic (FYM and Bio- Plus) and inorganic sources of nutrients in Meghalaya. The treatment consisted of three varieties *viz.*, Kufri Megha, Kufri Jyoti and Kufri Giriraj with three sources of nutrients, *viz.* 100% inorganic, 75% inorganic + 25% organic, 50% inorganic + 50% organic source of nutrients. The organic sources of nutrients were FYM and Bio- Plus. The experiment was laid down in a spit plot design with three replications.

Data presented in the Table 11.1 showed that the potato variety Kufri Giriraj recorded significantly highest yield (260.03 q/ha) over the varieties, Kufri Megha (224.70 q/ha) and Kufri Jyoti (207.95 q/ha). However, the potato varieties, Kufri Megha and Kufri Jyoti showed statistically at par yield. This might be due to varietal characters and early damage of Kufri Jyoti due to higher infestation of potato late blight.

The response of potato varieties to 100% inorganic recorded highest yield (254.43 q/ha) over 75% inorganic + 25% organic (220.52 q/ha) and 50% inorganic + 50% organic (208.72 q/ha). This might be due to mineralization of nitrogen in the early growth period of the crop. The response of potato varieties to different sources of organic nutrients *viz.*, FYM and Bio-Plus showed non significant effect.

Advanced studies on effect of inorganic and organic sources of nutrients on the yield of potato cultivars

On the basis of preliminary study, further experiment was conducted to analyze the effect of inorganic and organic (FYM) sources of nutrients on the yield of potato cultivars. The experiment consisted of three potato varieties, Kufri Jyoti, Kufri Giriraj and Kufri Megha with six sources of nutrient *viz.*, 75% inorganic + 25% organic NPK, 50% inorganic + 50% organic NPK, 25% inorganic + 75% organic NPK, 100% organic, 100% inorganic and control. The experiment was laid down in a split plot design with four replications.

Table 11.1 : Response of potato varieties to inorganic and organic sources of nutrients

	Treatments	*K. Megha*		*K. Jyoti*		*K. Giriraj*		
		FYM	*Bio-plus*	*FYM*	*Bio-plus*	*FYM*	*Bio-plus*	*Mean*
(*i*)	100% inorganic	243.00	246.60	240.5	248.6	308.55	226.40	254.43
(*ii*)	75% inorganic +25% organic	227.77	216.14	194.28	172.03	226.40	286.50	220.52
(*iii*)	50% inorganic + 50% organic	212.13	202.55	212.35	172.43	289.20	208.72	208.72
	Mean		224.70	207.95	260.03			
	CD (0.05)	Var	Source	Methods	S × M	V × M	V × M × S	
		33.45	NS	32.38	NS	NS	NS	

Table 11.2a : Response of potato varieties on growth and yield attributes to organic and inorganic sources of nutrient

Treatment	Varieties	Germination %	Plant Height (cm) 45 DAP	60 DAP	75 DAP	Number of Branches 45 DAP	60 DAP	75 DAP
1	*2*	*3*	*4*	*5*	*6*	*7*	*8*	*9*
75% Ing +25%Org NPK	Kufri Jyoti	84.33	23.57	32.81	37.80	5.40	5.70	5.79
50% Ing +50% OrgNPK		87.65	23.23	35.25	36.60	5.65	5.60	6.18
25% Ing +75% Org NPK		87.65	20.97	28.68	31.60	4.07	4.29	4.69
100% Org NPK		88.51	16.55	21.20	21.20	3.47	3.91-	4.20
100% Ing NPK		82.67	23.32	31.70	36.64	5.10	5.68	5.93
Control		83.50	17.78	25.90	29.64	3.62	4.00	4.20
75% Ing +25%Org NPK	Kufri Giriraj	85.17	25.37	32.36	39.47	4.00	4.85	5.56
50% Ing +50% Org NPK		82.67	24.09	37.10	38.47	4.35	4.71	5.38
25% Ing +75% Org NPK		91.13	21.80	34.93	35.60	4.53	4.90	4.94
100% Org NPK		89.35	15.94	26.20	30.80	4.40	4.75	5.20
100% Ing NPK		78.49	22.68	32.56	39.00	3.87	5.00	5.07
Control		79.61	12.90	22.50	29.47	3.40	3.75	4.20
75% Ing +25% Org NPK	Kufri Megha	88.51	23.11	36.60	40.14	3.64	4.79	4.90

(Contd...)

1	2	3	4	5	6	7	8	9
50% Ing +50% Org NPK		90.73	23.96	36.00	39.84	5.00	5.00	5.73
25% Ing +75% Org NPK		89.62	20.73	32.89	35.00	4.13	4.63	5.30
100% Org NPK		89.06	17.03	29.2	30.06	4.50	4.80	5.30
100% Ing NPK		89.30	20.71	30.93	34.60	4.36	5.20	5.38
Control		76.82	11.75	22.07	22.90	4.0	4.14	4.20

Table 11.2b : Response of potato varieties on yield to organic and inorganic sources of nutrient

Treat-ment	*Kufri Jyoti*		*Kufri Giriraj*		*Kufri Megha*	
	No. of Tubers/ha	*Yield ton/ha*	*No. of tubers/ha*	*Yield ton/ha*	*No. of tubers/ja*	*Yield ton/ha*
75% Inorg +25% Org Source of NPK	639583.33	19.30	694333.33	19.50	555555.56	21.15
50% Inorg +50% Org Source of NPK	673263.89	17.80	625000.00	17.60	322222.22	21.15
25% Inorgg +75% Org Source of NPK	684027.78	11.78	663402.78	16.40	556944.44	14.10
100% Org Source of NPK	693055.56	15.80	462250.00	13.90	481486.11	08.03
100% Inorg source of NPK	664472.22	20.60	568986.11	19.60	589347.22	18.00
Control	441394.44	11.67	316220.83	15.20	448152.18	12.00

Data presented in Table 11.2a showed that the average germination of potato varieties irrespective of sources were nearly same. However, the plant height and number of branches measured at different days interval were comparable in case of treatment 75% inorganic + 25% organic NPK, 50% inorganic + 50% organic NPK and 100% inorganic NPK but inferior plant height and number of branches were observed in the treatments 100% organic, control and 25% inorganic + 75% organic NPK irrespective of varieties.

The data presented in Table 11.2b revealed that highest average yield was recorded in treatments 100% inorganic NPK in Kufri Jyoti (20.60 t/ha) and Kufri Giriraj (19.60 t/ha) but in Kufri Megha, the highest yield (21.15 t/ha) was recorded in 75% inorganic + 25% organic NPK, 50% inorganic + 50% organic NPK. However, nearly equal average yield were recorded in treatments 75% inorganic + 25% organic NPK, 50% inorganic + 50% organic NPK and 100% inorganic NPK in each variety separately. The average yield of potato varieties were found poor in treatments 100% organic, control and 25% inorganic + 75% organic NPK irrespective of varieties.

Effect of organic and inorganic sources of nutrients on seedling tuber (92-PT-27) and minituber field generation II (Kufri Giriraj)

An experiment was conducted during summer to study the effect of organic and in-organic sources of nutrient on seedling tuber (92-PT-27) and mini tuber field generation–II (Kufri Giriraj). The treatments consisted of two cultivars V1 = Kufri Giriraj (Mini tuber FG-II) and V2 = TPS (92 PT-27) and six treatments T1 = Recommended dose of fertilizer, T2 = Recommended dose of fertilizer + Mycorrhiza, T3 = Recommended dose of fertilizer (75%) + Bio-plus (25%), T4 = Recommended dose of fertilizer (75%) + FYM (25%), T5 = Recommended dose of fertilizer (50%) + Bio-plus (50%) and T6 = Recommended dose of fertilizer (50%) + FYM (50%). The experiment was laid down in a randomized block design (RBD) with three replications.

Data presented in Table 11.3 reveals that highest plant height and number of branches in TPS (92 PT-27) were statistically significant over Kufri Giriraj (FG–II) irrespective of different treatments. On the other hand, highest plant height

Table 11.3 : Effect of organic and inorganic sources of nutrients on seedling tuber (92-PT-27) and minituber field generation II (K. Giriraj)

Variety/ TPS	Treatments	Germi-nation	Plant height (cm)		Number of branches		Yield (t/ha)	No. of Tuber
			45 DAP	75 DAP	45 DAP	75 DAP		
K. Giriraj	V_1T_1	96.20	49.13	59.67	2.40	3.10	16.92	352078
	V_1T_2	95.40	40.67	52.07	2.40	3.00	20.83	456243
	V_1T_3	93.80	45.60	48.40	2.10	2.60	14.58	320828
	V_1T_4	90.00	39.93	49.40	2.60	2.60	12.29	302078
	V_1T_5	91.20	38.00	46.20	2.10	2.30	13.12	349349
	V_1T_6	94.00	37.67	51.87	2.20	2.90	11.46	304162
				Average yield			14.90	
92-PT-27	V_2T_1	95.20	44.67	72.53	3.70	3.10	14.37	344786
	V_2T_2	96.00	49.40	68.58	4.00	3.20	13.54	339578
	V_2T_3	94.00	46.67	64.47	3.30	3.00	11.46	236871
	V_2T_4	93.20	46.86	56.40	2.70	3.10	10.42	289579
	V_2T_5	94.50	40.33	58.13	2.40	2.70	9.37	237496
	V_2T_6	90.00	39.60	59.87	2.90	2.90	12.79	293745
				Average yield			12.30	
$CD_{0.05}$		NS	CD (V) 0.93 CD (T) 0.65	CD (V) 3.82 CD (T) 6.61	CD (V) 0.40 CD (T) NS	CD (V) 0.37 CD (T) NS	CD (V) 1.20 CD (T) 2.10	CD (V) 34499 CD (T) 59782

V1 = K.Giriraj and V2 = 92-PT-27

T1 = RDF,T2 = RDF + Mycorrhiza,T3 = RDF(75%) + Bio-plus(25%), T4 = RDF(75%) + FYM(25%),T5 = RDF(50%) + Bio-plus(50%) and T6 = RDF(50%) + FYM(50%) RDF = Recommended dose of fertilizer.

and number of branches recorded in Kufri Giriraj (FG-II) in the treatment T1 = Recommended dose of fertilizer (59.67 cm and 3.10) followed by T2 = Recommended dose of fertilizer + Mycorrhiza (52.07 cm and 3.00) but in TPS (92 PT-27) highest plant height was recorded in T1 = Recommended dose of fertilizer (72.53 cm) which is at par with T2 = Recommended dose of fertilizer + Mycorrhiza and highest number of branches were recorded in T2 = Recommended dose of fertilizer + Mycorrhiza (3.2) though it was statistically non significant.

The highest yield was recorded in Kufri Giriraj (14.90 t/ha) which was statistically significant over TPS (92 PT- 27) which recorded 12.30 t/ha. In cultivar Kufri Giriraj, the treatment T2 = Recommended dose of fertilizer + Mycorrhiza gave highest yield (20.83 t/ha). However, T1 = Recommended dose of fertilizer and T3 = Recommended dose of fertilizer (75%) + Bio- Plus (25%) showed at par yields. In cultivar TPS (92 PT- 27), highest yield was recorded in the treatment T1 = Recommended dose of fertilizer (14.37 t/ha) which was at par with treatment T2 = Recommended dose of fertilizer + Mycorrhiza which followed and T3 = Recommended dose of fertilizer (75%) + Bio-Plus (25 %) that recorded 13.54 t/ha and 11.46 t/ha respectively.

CHAPTER 12

Evaluation of Fungicides/Fungicide Combinations Against Late Blight Disease in Autumn TPS Grown Crop

Potato is grown throughout the year as summer, spring and autumn crop in hilly areas of India. Autumn TPS grown crop has been recognized as an important means to produce quality seed in Meghalaya hill condition for its subsequent utilization in the summer season (Kumar *et al.*, 2003). Late blight caused by Oomycetes pathogen *Phytophthora infestans* (Mont.) de Bary is one of the devastating disease of potato worldwide and has become a serious threat to potato cultivation with the breakdown of resistance in commercially cultivated varieties in most of the countries. In India, losses due to the disease may go as high as 85 per cent in the hilly regions, if the crop (susceptible cvs.) remains unprotected. Many fungicides have been reported effective and recommended against potato late blight (Dutt *et al.*, 1982; Schwin and Morgot, 1991). However, due to high disease pressure in Meghalaya hill, particularly in the autumn season, when inoculum load is very high in the moist and humid weather, most of the recommended fungicides often fail to provide adequate control. Systemic fungicides alone would fail to achieve effective control due to development of resistant strains of the pathogen (Kadish and Cohen, 1988). Therefore, study was undertaken to evaluate four fungicide combinations including systemic and contact fungicides for effective management of late blight disease of potato in autumn TPS grown crop in Meghalaya hills.

Field experiments were conducted during autumn season (July to December) with TPS grown crop (hybrid 92-PT-27). The TPS seedlings were raised in nursery beds prepared inside the green house and transplanted to the main field in the second week of August in both the years. All recommended practices like, fertilization (10 tones/ha FYM and N-P-K @ 120-120-60 Kg/ha), weeding and twice earthing-up were done timely and adequately. Three commercial fungicides namely Dithane M-45 (mancozeb 75% WP @ 0.2%), Indofil M-45 (mancozeb 75% WP @ 0.2%) and Ridomil MZ-72 (metalaxyl 8% + mancozeb 64% WP @ 0.25%) were tested alone and in different combinations. The experiment was conducted in randomized block design and each treatment was replicated four times. The plot size kept was 3.20 m ´ 3.00 m consisting of 8 rows and 20 plants/row at a spacing of 40 cm ´ 15 cm. Plain water sprayed plots were kept as control for comparison. The crop was sprayed just before the normal appearance of blight i.e. 1st week of September with high volume sprayer (700-800 lit/ha). Systemic fungicides were sprayed at 15 days interval and spray of contact fungicides was done at 8 days interval. Data on disease severity was recorded at weekly intervals according to British Mycological Society (BMS) scale (Anonymous, 1947), starting from the appearance of the disease. Area under disease progress curve (AUDPC), a method of measurement of disease progress was calculated (Abbot, 1973) for each replication to compare different treatments and were analyzed statistically. Delay in onset of disease was calculated over check plots. Yield data were also recorded and calculated on hectare basis.

Late blight appeared on the crop in the second week of September and 100 per cent crop damage occurred in control plots at the end of October. Two years pooled data (Table 12.1) revealed that various treatments reduced late blight severity to varying extent compared with that in unsprayed plots, and these results indicated the critical importance of delaying the onset of disease to prevent losses in yield. Prophylactic spray of Ridomil MZ-72 delayed the onset of disease by 6.5 days followed by Indofil M-45 (3.0 days) and Dithane M-45 (2.5days).

All the treatments significantly reduced AUDPC and increased the yield. The trend in reduction of AUDPC and increase in yield in different treatment was almost same in both the years under study as there was no significant difference between year and treatment. AUDPC analysis showed significant difference in disease progress in different fungicides. The blighting rate (AUDPC) was least (83.95) in the treatment consisting of two sprays of Ridomil MZ-72 plus three sprays of Indofil M-45 (T_5) in the sequence of systemic, contact, systemic, contact and contact. This was followed by two sprays of Ridomil MZ-72 plus three sprays of Dithane M-45 (T_4). However, both the treatments were rated at par in checking the blighting rate. Treatments T_4 and T_5 were statistically superior over the treatments consisting of five sprays of Indofil M-45 (T_3) and five sprays of Dithane M-45 (T_2) alone. Earlier workers have also reported the effectiveness of Ridomil MZ against *Phytophthora infestans* in laboratory and field test (Singh and Sharma, 1993; Singh *et al.*, 1997). The result might be due to the fact that metalaxyl reduced both growth and sporulation of the pathogen whereas mancozeb inhibits spore germination only (Bichn *et. al.*, 1978). No significant difference was found between Indofil M-45 and Dithane M-45 in reducing the AUDPC.

Table 12.1 : Effect of different fungicides on late blight infection and yield of potato

Treatment	*Delay in disease onset (days)*	*AUDPC*	*Yield (tonnes/ ha)*	*% yield advantage over control (%)*
T1 (Unsprayed control)	–	414.46	51.95	–
T2 (5 sprays of Dithane M-45)	2.5	134.80	112.76	117.05
T3 (5 sprays of Indofil M-45)	3.0	125.61	124.23	139.13
T4 (2 sprays of Ridomil MZ-72 plus 3 sprays of Dithane M-45)	6.0	99.19	139.91	169.32
T5 (2 sprays of Ridomil MZ-72 plus 3 sprays of Indofil M-45)	6.5	83.95	146.77	182.52
CD (P = 0.05) Treatment		19.86	11.49	
Year × treatment		NS	NS	

The data on tuber yield also indicated that all the test fungicides gave significant increase in yield over control. Plots receiving two sprays of Ridomil MZ-72 plus three sprays of Indofil M-45 (T_5) recorded maximum yield (146.77 q/ha). Two sprays of Ridomil MZ-72 plus three sprays of Dithane M-45 (T_4) gave result at par with T_5. The treatments consisting alternate sprays of metalaxyl and mancozeb (as in case of T_4 and T_5) gave significantly higher yield over the treatments having spray with mancozeb alone (as in case of T_2 and T_3). Metalaxyl sprays were found effective against late blight and they gave significant increase in yield because of the properties of remaining effective on foliage for 18 days (Bhattacharyya *et. al.*, 1983).

The present study confirmed that alternate spray of Ridomil MZ-72 @ 0.25% at 15 days interval along with mancozeb @ 0.2% at 10 days interval would be useful for effective and economic management of late blight and minimize the yield losses to TPS grown potato crop due to the disease in the autumn season in Meghalaya hills. However, with the advert of rew molecules of systemic fungicide, tere is greater scope for such expesimentations to establish and confirm the above observations.

CHAPTER 13

Potato Late Blight and its Integrated Management Under Meghalaya Hills

Late blight of potato is primarily a foliar disease but it readily infects the tubers. Therefore, effective management of the disease would entail reduction of both foliar and tuber infection. The total amount of fungicides applied annually on potato is greater than for any other crops. About 97 percent of potato hectarage is treated with fungicides. The cost of chemical control of late blight worldwide may well exceed above 1 billion annually, although actual losses in the field and storage are likely to be several fold higher. Many fungicides have been reported effective against and recommended to control the disease, but most of the fungicides often fail to provide adequate control if not applied at appropriate time. Therefore, effective management can be achieved by employing chemicals, cultural practices and host resistance in an integrated manner. Use of naturally occurring bio-control agents is considered as an alternative of fungicides. Antagonism to *P. infestans* by some naturally occurring microorganism viz., *Trichoderma viride*, *Pseudomonas syringae*, *Fusarium graminearum*, *Pythium ultimum* etc has been reported. Various integrated disease management (IDM) components have been evaluated mostly individually against potato late blight. There are only a few studies wherein all the resistant components have been used together in an integrated manner.

Considering the above background, the present investigation is undertaken with the objectives namely i) Study on collection of isolates of *Phytophthora infestans* and identification of mating types and physiological races of the pathogen, ii) Study on foliar aggressiveness of different mating types and different isolates within the same mating type of *Phytophthora infestans* on detached leaflets of potato cultivars (susceptible, moderately resistant and resistant cultivars) and

iii) Study on integrated management of *Phytophthora infestans*, the incitant of late blight disease of potato.

Table 13.1 : Intensity of late blight at weekly interval after appearance of the disease

Treatments	*Disease intensity*							*Delay in disease onset over inoculated control (days)*
	7 days	*14 days*	*21 days*	*28 days*	*35 days*	*42 days*	*49 days*	
T_1	4.00	16.25	47.50	88.75	96.25	100.00	100.00	1.50
T_2	6.25	21.25	52.50	91.25	97.50	100.00	100.00	-
T_3	0.53	3.00	5.25	10.00	15.00	21.25	28.75	6.25
T_4	1.50	10.00	38.75	80.00	90.00	96.25	100.00	3.75
T_5	0.33	2.50	5.00	9.50	15.00	16.25	25.00	6.75
T_6	0	0	0	0.78	1.00	2.25	4.00	23.75
T_7	0	0	0	1.50	3.50	5.00	7.75	24.75
T_8	0	0	0.25	1.00	1.75	2.75	4.50	22.50
T_9	0	0	1.00	3.00	4.50	6.25	8.50	18.00
T_{10}	0	0	1.00	3.00	4.75	6.25	7.75	17.00

Figures are the average of four replications.

Following Treatment combinations *viz.*, T_1: Susceptible cultivar (Kufri Jyoti), T_2: Susceptible cv + *Phytophthora infestans*, T_3: Susceptible cv + *P.infestans* + Chemical (best fungicide'! metalaxyl '! best fungicide '!metalaxyl) (four sprays), T_4: Susceptible cv + *P.infestans* + Bio-control (best antagonist) (spore suspension four sprays), T_5: Susceptible cv + *P.infestans* + Chemical + Bio-control (best fungicide + best antagonist '! metalaxyl + best antagonist '! best fungicide + best antagonist '! metalaxyl + best antagonist) (four sprays in combination), T_6: Resistant cv + *P.infestans* + Chemical + Bio-control (best fungicide + best antagonist '! best fungicide + best antagonist) (two sprays in combination), T_7: Resistant cv + *P.infestans* + Bio-control (best antagonist) (spore suspension two sprays), T_8: Resistant cv + *P.infestans* + Chemical (best fungicide '! best fungicide) (two sprays), T_9: Resistant cv + *P.infestans* and T_{10}: Resistant cv (Kufri Megha) were taken with each of 4 (Four) replications on the plot of size: 3.6 m × 3.0 m with spacing: 60 cm × 20 cm during 2005-06. The findings are presented in the Tables 13.1 and 13.2 and described as follows:

Table 13.2 : Yield of tuber at the time of harvest

Treatments	*Yield of tuber (kg / plot)*					*Yield (q/ha)*
	Large	*Medium*	*Small*	*Rottage*	*Total*	
T_1	1.10	3.55	3.55	1.13	9.33	86.39
T_2	1.08	2.41	3.84	1.34	8.66	80.19
T_3	6.99	9.25	4.55	0.16	20.95	193.98
T_4	2.55	3.64	3.41	1.06	10.66	98.70
T_5	7.11	10.66	4.10	0.15	22.03	203.98
T_6	6.80	15.43	4.99	0.05	27.26	252.41
T_7	5.43	13.51	5.75	0.09	24.78	229.44
T_8	6.75	14.58	5.00	0.08	26.40	244.44
T_9	5.19	11.10	5.05	0.18	21.51	199.17
T_{10}	5.21	11.68	5.05	0.16	22.10	204.63

Figures are the average of four replications

From the mean values (Table 13.1), it can be seen that the minimum disease intensity (4.00) was recorded in the treatment T_6 at 49 days after first disease appearance. This was followed by the treatment T_8 (4.50) and T_7 and T_{10} (7.75). Whereas, the disease intensity increased very rapidly and all crops were died at 42 days after disease appearance in the treatments T_1 and T_2 comprising no chemical or biological treatment with susceptible cultivar (Kufri Jyoti). The treatment consisting biological control alone in susceptible variety (T_4) showed less disease intensity at preliminary stage (up to 21 days), but finally all crops died. Similarly, mean yield of tuber were calculated out for each treatment and presented (Table 5.13). The data reveals that the highest total yield of tuber (27.26 kg/plot) was recorded with the treatment T_6 consisting chemical and biological control along with host resistance. The same treatment also recoded lowest quantity of rottage tuber (0.05 kg/plot). Treatments T_8 and T_7 were also as effective as T_6 in producing total tuber yield (26.40 and 24.78 kg/plot respectively) and yield of rottage tubers (0.08 and 0.09 kg /plot respectively). The treatment with no control measure (T_2) produced the lowest total yield of tuber (8.66 kg/plot) and highest yield of rottage tubers (1.34 kg/ plot).

PART - 4

TRANSFERRABLE TECHNOLOGIES AND ON-FARM ADAPTIVE RESEARCHES

CHAPTER 14

RECOMMENDED POTATO VARIETIES

Potato was introduced into India in the beginning of the 17th Century and was reported to have been cultivated in several parts of the country including the North Eastern parts. However, records show that potato cultivation probably came into existence with the introduction of popular variety '*Khasi*' in Meghalaya (then Assam hills) around 1830. '*Khasi*' proved susceptible to late blight and a number of other varieties were introduced from England and Australia between 1892 and 1946 and evaluated against some local varieties. However, in the subsequent trials *Arran Consul*, *Kerr's Pink*, *Royal Kidney*, *Up-to-Date* and *Great Scot* showed promise and hence these were multiplied for further distribution. These varieties were named locally and continued to be commercially cultivated till about the mid and late seventies and after wards when improved and high yielding varieties like Kufri Jyoti and K. Khasigaro were introduced in the state.

The states like Assam, Meghalaya and Tripura contribute to about 81% of total area and account for 87% of the total production of NE region (Fig 14.1 and Fig 14.2). A number of factors have been identified that contribute to the poor productivity in the region. Major factors among them are rainfed conditions, heavy incidence of pest and diseases (late blight, bacterial wilt and, wart in Sikkim), inadequate availability of inputs like, fertilizer, plant protection chemicals, quality seed of high yielding varieties and continuation of traditional practices of potato cultivation as discussed in this book in previous chapters also.

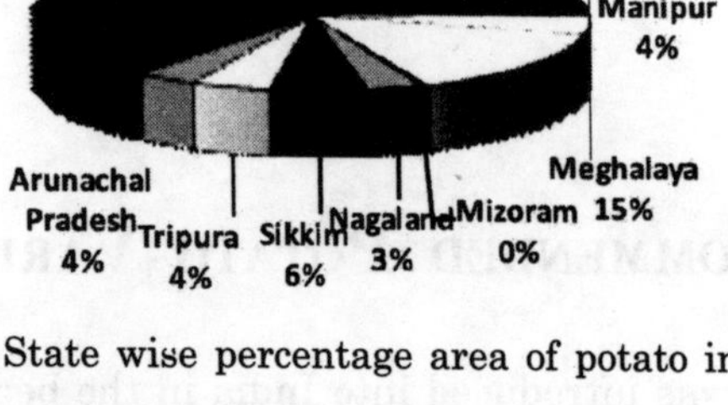

Fig. 14.1 : State wise percentage area of potato in NE region

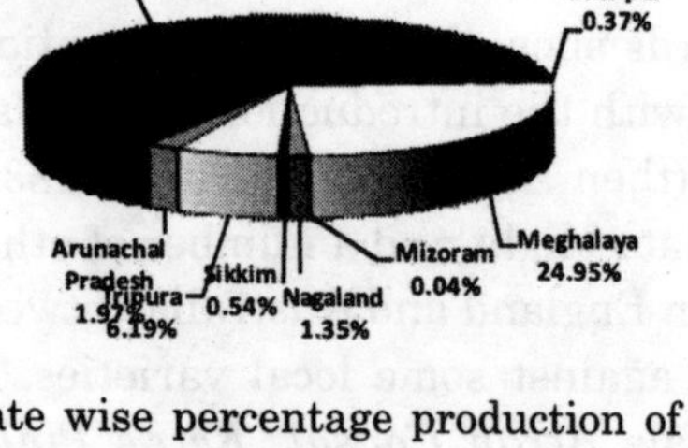

Fig. 14.2 : State wise percentage production of potato in NE region

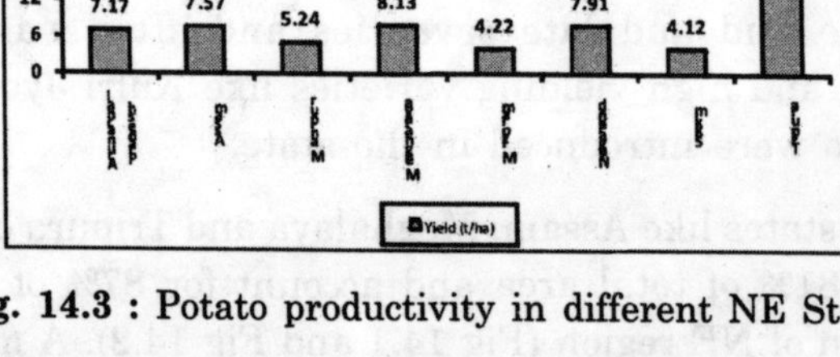

Fig. 14.3 : Potato productivity in different NE States

Source: Directorate of Economics and Statistics, Ministry of Agriculture Govt. of India.

Production Agro-ecology

Potato crop is grown as a cash crop round the year in the North Eastern Region. In the hilly areas of the region under rainfed conditions, potato crop is taken as summer crop (Feb-March to July-August), as well as autumn crop (August to December). It is grown as winter crop in plain of Assam,

Tripura, Manipur valley, lower parts of Arunachal Pradesh and Dimapur District of Nagaland during October to February; where as, in Sikkim potatoes are grown in the spring season (Jan/Feb-July/Aug.).

Altitude	*Planting time*	*Harvesting time*	*Parts of NE Region*	*Remarks*
Plains (up to 500 m)	Sept- Oct	Jan- Feb	Assam, Tripura, Manipur valley, Dimapur district of Nagaland and lower parts of Arunachal Pradesh	Irrigation is necessary for good crop growth
Mid-hill (upto1500 m)	Feb.-March	End June to end July	Siang district (Basar area) of Arunachal Pradesh, Jowai sub division of Jaintia Hills, some areas in Khasi Hills	Rainfed crop
High hills (up to 2400 m)	March-April	July-August	Parts of Arunachal Pradesh, Upper Shillong, Mawphlang and Mairang area of Meghalaya, Mao and Ukhrul area of Manipur, Blue mountains in Mizoram	Rainfed crop
Very high hills (above 2400 m)	-do-	-do-	Tawang, Upper Subansiri, Siang and Kameng districts of Arunachal Pradesh	Rainfed crop

High Yielding Potato Cultivars for Suitable NE Region

Central Potato Research Station, Shillong, a regional station of Central Potato Research Institute, Shimla has been engaged in conducting researches on the ways to mitigate the problems of potato production in North Eastern region since 1959. In this process the station has released/introduced several improved potato cultivars, viz. Kufri Khasigaro, Kufri Jyoti, Kufri Megha, Kufri Giriraj, Kufri Himalini and Kufri Girdhari suitable for cultivation in the North Eastern region. Kufri

Khasigaro released in 1968 is no longer in cultivation because of the tuber shape, poor keeping quality and short dormancy. Kufri Jyoti released in 1968 occupies the largest area in this region. Though, Kufri Jyoti has now lost resistance to late blight resulting in its yield reduction, the variety is still popular because of its fast bulking habit, good keeping quality and taste. Among others, Kufri Megha released in 1989 is also under cultivation, though it is a slow bulker but has high resistance to late blight, hence gives assured income to the farmers. Kufri Giriraj released in the year 1998 is gaining popularity in this region because of high yield and resistance to late blight. Farmers are, however, sceptical about its poor keeping quality and short dormancy. Its short dormancy is being exploited by using the produce of early summer crop as seed for the planting the following autumn potato crop. The cultivar, Kufri Kanchan with resistance to late blight and wart was released in 1999 for North Bengal hills and Sikkim. This variety is also becoming choice of the farming community because of preference for red skin tubers. Recently released varieties, Kufri Himalini and Kufri Girdhari have performed well in on-farm trials and demonstrations and gaining popularity because of late blight resistance and attractive tuber attributes. North Eastern region has been identified as one of the potential region favoring accumulation of high tuber dry matter. In this context, this region can play an important role for ensuring round the year availability of processing grade potatoes for up coming potato processing industries of the region. The recently released potato variety Kufri Himsona for processing purpose suitable for cultivation in hills has also shown promise in the trials conducted in Shillong.

In the plains agro-ecosystem as prevailing in Assam, Tripura, Manipur valley, Dimapur district of Nagaland and lower parts of Arunachal Pradesh varieties like K. Badshah, K. Pukhraj are being grown in some of the potato growing areas. However, K. Jyoti still covers maximum area under potato due to non-availability of the quality seed of said varieties.

Use of TPS (True potato seed) is becoming popular for generating low cost quality tuberlets for planting material as well as for table use. TPS can be utilized either by raising seedling transplant in the nursery and subsequent transplanting in the field in the same season or by production of seedling tubers (G_0 generation) in the nursery for raising commercial potato crop in the next season. The produce obtained from TPS can be retained as seed for another 3-4 years by following strict control regime against aphid, brown rot and late blight.

Thus, TPS could be a viable solution to mitigate the problems of non-availability of quality seed potato in this region. Moreover, breeder seed production cannot be undertaken in the mid hills and low land areas occupying more area under potato than the high hills. In addition, most of the potato farmers in this region are small and marginal. Hence, TPS could be an appropriate technology for them due to the low cost of cultivation. The states like Meghalaya, Manipur, Nagaland and Arunachal Pradesh grow potato in two seasons. Hence, seedling tubers produced in one season could be stored easily under ambient conditions in diffused light and utilized for ware potato production in the following crop season. Thus the need for costly cold stores could be dispensed by the use of TPS in these states. In the states like Assam and Tripura where temperatures are mild in winters, potato crop could be produced by transplanting seedlings or planting seedling tubers produced in the preceding season and stored in the cold store. 5-10 q seedling tubers required for planting in a hectare, would need less storage space in the cold store as compared to nearly 30 q of seed potatoes needed to plant the same area. The characteristic features of suitable cultivars/ TPS population for the region are described below-

(*i*) Kufri Jyoti (1968)

Plant Characteristics : Medium to Early maturing, plants are tall, erect, compact and vigorous with few but thick stems. The flowers are white. Tubers are medium to large in size, oval in shape with white skin and fleet eyes. The sprouts are blue- purple.

Yield Potential : 20-30 t/ha

Disease : Moderately resistance to late and early blight. Slow rate of degeneration.

Cooking and Processing Quality : Easy to cook, waxy texture, mild flavour. Suitable for processing

(*ii*) Kufri Megha (1989)

Plant Characteristics : Late maturing, plants are medium tall, semi-erect, medium compact with green and straight stems. The flowers are white. Tubers are white, smooth surface and are oval in shape with shallow eyes. The sprouts are dull green.

Yield Potential : 22-28 t/ ha

Disease : Highly resistant to late blight.

Cooking and processing quality : Good cooking quality, floury texture, Not suitable for processing

(*iii*) Kufri Giriraj (1998)

Plant Characteristics : Medium maturing variety, plants are medium tall, semi-erect with medium thick stems that are coloured at the base. The flowers are light purple. Tubers are white with smooth skin, medium to large in size, oval in shape with white flesh, and fleet eyes. The sprouts are light purple.

Yield Potential : 25-30 t/ ha

Disease : Resistant to late blight.

Cooking and Processing Quality : Easy to cook, waxy texture, not suitable for processing

(*iv*) Kufri Kanchan (1999)

Plant Characteristics : A Medium maturing variety, plants are medium tall, semi-erect, medium compact, vigorous with red purple flowers. Tubers are red medium to large in size, oval oblong with fleet eyes. The sprouts are blue purple.

Yield Potential : 25-28 t/ ha

Disease : Field resistance to late blight, resistant to Wart and slow rate of degeneration.

Cooking and processing quality : Easy to cook, waxy texture, not suitable for processing

(v) Kufri Himalini (2006)

Plant Characteristics : A medium maturing variety producing medium-tall, semi-erect, vigorous, semi-compact plants with light purple flowers. The tubers are medium, oval-oblong, white, smooth, shallow eyes with pale yellow flesh. Sprouts are light purple in colour.

Yield Potential : 30-35 t/ ha

Disease : Field resistance to late blight in foliage and moderate resistance in tubers.

Cooking and Processing Quality : Easy to cook, waxy texture.

(vi) Kufri Girdhari (2008)

Plant Characteristics : A medium maturing variety producing medium-tall, semi-erect, vigorous, semi-compact plants with white flowers. The tubers are medium, oval-oblong, white, smooth, shallow eyes with pale yellow flesh. Sprouts are purple in colour.

Yield Potential : 27-30 t/ ha

Disease : High level of field resistance to late blight in foliage and tubers.

Cooking and Processing Quality : Easy to cook, Mealy texture.

(vii) Kufri Himsona (2006)

Plant Characteristics : A medium maturing (120-140 days) variety producing open, medium, thin stems with flowers primarily purple/intense purple with white tip on both sides. The tubers are medium, round-oval, white, smooth, shallow eyes with cream pale yellow flesh. Sprouts are purple in colour. It is the 1st high dry matter, low reducing sugar potato variety.

Yield Potential : 20-25 t/ ha

Disease : Resistant to late blight.

Cooking and Processing Quality : Easy to cook, floury texture, suitable for processing

(viii) TPS Population 92-PT-27 (2007)

Plant Characteristics : Uniform and high yielding TPS population. Both parents flower under short days in the plains. Suitable for cultivation in eastern region

Yield Potential : 18-20 t/ ha in plain areas and 5.0-8.0 t/ ha in hills

Disease : Resistant to late blight.

Cooking and Processing Quality : Suitable for table purpose.

North eastern regions are suitable for growing potato round the year in one or another part and therefore, varieties may be developed according to local needs and preference of the peoples for making availability of potato as main part of diet and improving the economic conditions of the farmers. Varieties with improved physical/nutritional qualities, good keeping quality, processing attributes, high degree of consumer preference, improved disease resistance (late blight and bacterial wilt) are required to meet the farmer demand. Varieties for specific microclimatic conditions for the region like low lying areas of hills which need early bulking varieties with moderate levels of late blight resistance.

CHAPTER 15

Technology for Quality Seed Potato Production

Prevailing Seed Production Systems

The farmers of North Eastern Region rely on the seed supplied through formal seed system as well as largely through informal seed system that is produced traditionally or purchased locally without keeping in mind the various aspects essential for quality seed production.

Informal Seed Production / Source

Potato farmers of North Eastern states especially hilly region use to multiply their own seed. During the production process little measures are taken to safe guard the seed quality. However, the areas situated in valley / low hills rely on the seed supplied by the local traders or from states near by. However if the farmers are quality conscious and take proper measures to ensure good quality seed multiplication for their own use may be a welcome step. It will reduce the cost on the production of certified seed each year.

Formal Seed Production / Source

Formal seed production system combines the use of techniques that are appropriate for the production of high quality seed with an organizational and legal system that verifies and certifies the quality of product. Central Potato Research institute and its various regional station are involved in production/multiplication and distribution of nucleus / breeder and foundation seed to various state Govt. for the production of foundation and certified seed in the country including NEH.

Development of Nucleus Seed

The nucleus seed is produced at five stations of CPRI in plains and hills. Nucleus seed can be developed either through clonal selection and tuber indexing or through meristem tip culture and micro propagation.

Clonal Selection

Apparently healthy looking plants are selected in the field based on visual inspection, marked and pegged. 75-80 such plants are sufficient to feed the seed requirement of 1 ha in forth year

Tuber Indexing

Single eye from each of the four-selected tuber is scooped from rose end and planted in glass house in 10 cm pots. Counterparts are cement dusted and retained for field planting on their being declared healthy. Each emerged plant let is tested after 6-7 weeks of planting at 6-8 leaf stage through ELISA against PVX, S, M, Y & A. A single positive reaction in any one of these tested results in rejection of all the four tubers indexed.

Multiplication of Indexed Material

The disease free indexed tubers are multiplied in the field in **stage-I** at a spacing of 1 × 1m distance. The produce of stage-I is further multiplied in **stage-II** at a spacing of 1 × 0.2m or 1.2 × 0.2m in hills and plains respectively. The seed produced in these stages is called as Nucleus Seed.

Production of Breeders' Seed

The produce of nucleus seed (stage-II) is multiplied in 3rd year as pre basic seed in **stage-III** at a normal spacing of 60× 20cm. This is further multiplied in 4th year in **stage-IV**. The produce of fourth year multiplication is called breeders seed of potato.

Cultural practices during breeders seed production

Seed stocks should be pre-sprouted before planting which helps in uniform, quick and full germination. For pre-sprouting seed stocks should be removed from cold store 7-15 days before planting keep the stocks in pre cooling chamber of cold store for 24 hours. Spread these in shade or in cool place for sprouting. It is better to fill the seed stocks in chitting trays of 75x 45 × 15cm size made of wooden splints or plastic. Proper combination of seed size and spacing is essential to get required number of stems per hectare. Spacing for different size of seed tubers may be 25-50g at 50 × 20cm or 60 × 15cm; 50-100g at 60 × 25cm and 100g and above at 60 × 30 or 60 × 40cm. The fertilizer doses for plains should be 100 kg nitrogen; 60 kg phosphate and 120 kg of potash per hectare and 50 kg nitrogen at earthing up. In hills, 100:100:100, N, P, K ha may be applied alongwith20 kg, N/ha at earthing up.

Field testing and visual inspection

During testing, if a single plant is found infected with any of the viruses under test the entire clone is rogued out along with its produce. Based on the ELISA results, the healthy counterpart tubers are selected and planted in the nucleus stage-I for further multiplication. Three roguings at 45, 60 and 75 days after planting also undertaken to remove diseased, off type and bolter plants.

Stage-I : Each plant is tested individually through ELISA for viruses X, S, Y, M and A. The diseased plants are rogued out along with its produce, if any.

Stage-II : Serological testing is carried out for viruses × & S through composite sample of the three plants taking two leaves from each plant. Other viruses/phytoplasmas are observed by visual inspection. In case of infection, entire clone has to be rogued out.

Stage-III : Incidence of viruses × & S is determined through random check test by serologically testing 300 plants/ha for PVX + S. While for other viral/phytoplasmal disease visual observations are made.

Stage-IV : The virus-free nucleus stocks are subsequently multiplied in stage-IV and its produce is called breeder's seed. The incidence of viruses × & S is determined by serologically testing of 150 plants/ha, through random check test.

Foundation Seed (FS–I & FS–II)

The two types of seed are produced from breeders and foundation seed respectively by state Agriculture/Horticulture departments & National Seeds Corporation. Among North-Eastern states Meghalaya, Manipur, Mizoram, Nagaland, Arunachal Pradesh & Sikkim are involved for production of these categories of seed in hills under long day condition.

Certified Seed

It is a further multiplication of foundation-II seed at registered growers field by following the package of practices for seed plot technique in the areas suitable for quality seed production. The FS-I & II seed is inspected and certified by State Seed Certification Agencies.

Quality Seed Production

Significance of Quality Seed Potato

In successful potato cultivation seed is a critical input. It accounts to about 50 % of the total cost of potato cultivation. The crop is input intensive and it is affected seriously by the use of quality seed, irrigation and balanced nutrition. If poor quality seed tubers are used then no other input can help in improving the yield. The use of quality seed is not only for high production and for quality production, but also for sustainable production system. Quality seed potato will give the crop a good start. Whereas the negative effects of using are difficult to restore later on in the subsequent seasons. Potato is traditionally grown through seed tubers, it is liable to get systemic infection of viral and mycoplasma and several other surface born fungal, bacterial and nematode diseases. The viruses are systemic in nature and transmitted by various modes including contacts, aphids and leaf- hoppers, survive in the tubers year after year and cause degeneration in subsequent generations. Normally, viral infections do not show any symptoms on potato tubers, but are carried through the tubers and cause yield reduction in the subsequent crops. Several studies have also supported this fact. Pushkarnath (1959) reported a yield loss to the extent of 88.5 percent and

Khurana and Singh (1988) reported a yield loss of 47.0 to 61.0 percent within three years if seed was not replaced with healthy high yielding cultivars. There is no direct control measure at field levels to keep the build-up of these diseases within limits in the crops. These biotic stresses affect the plant stand, vigour, production potential and ultimately the crop yields. Therefore, to maintain the crop productivity and achieve higher yields, it is essential to have a workable system through which purity, cleanliness and health of seed is maintained. The control of the insects indirectly controls the spread of viruses. Simultaneously elimination of virus infection foci through the process of detection either through serological means or visual inspection and rouging the disease plants further strengthen the process of quality seed production.

Procedures for Quality Seed Production

Suitable Areas

A quality seed can be produced in areas and fields, which are free from serious soil borne pathogens and pests.

Suitable Varieties

Keeping in mind the two seasons for seed production existing in NEH region following varieties have been recommended.

Varieties	*Characteristics*
Kufri Giriraj	Medium maturing 130-135 days, resistant to late blight both foliage and tuber, yield potential 25 tonnes per hectare, dormancy-medium (8-9 weeks)
Kufri Jyoti	Hills: Medium to early maturing 110-130 days (in plains, 90-100 days), Moderately resistant to late and early blight, resistant to wart, slow rate of degeneration.
Kufri Megha	Medium to late maturing, highly resistant to late blight,
Kufri Kanchan	Red skin variety, resistant against wart and late blight disease, medium maturing (135 days)

Seed Preparation

The seed stocks should be taken out in the last week of Feb in the hills and last week of September in the valley / low hills. The seed tubers are kept in diffused light either in wooden / plastic crates or by spreading on the floors. The chitting starts with the apical bud. But due to apical dominance the other bud remain dormant. The diffused light helps in greening the apical sprout with restricted growth due to formation of abcisic acid. The process stimulates the sprouting of lateral buds. Thus, more number of sprouts/tuber, which, will help in getting good yields. Presprouting is also helpful in removing the hairy, thin sprouted or blind tubers, thus avoiding gap in germination. Before planting potato tubers for quality seed production following aspects need to be checked:

(i) Health

The presence of disease and pest in seed not only affect the growing capacity/yield of the crop but also has negative effects on the marketing / consumption quality of the tuber. Use of diseased/infected tuber (bacterial wilt), will help in spreading diseases from one area to others they did not occur before.

(ii) Physiological Age

After harvest, potato tuber goes through different stages of development, which have their effect on the sprouting capacity and growth vigour. They are influenced by internal (genetical) and external (environmental) conditions, like temperature, humidity, light and chemical agents. They can be manipulated to a considerable extent.

The successive physiological stages that tuber goes through are : (*a*) dormancy-no sprout, (*b*) apical dominance-only one sprout will develop, (*c*) normal sprouting stage-several vigorous sprouts will grow, later these sprouts will produce branches and (*d*) senility-tubers shrivel, hair sprouts will be formed, low growth vigour, and 'little potato' growth.

Depending on the variety and the conditions, the tubers will take a longer or shorter time to go through these stages.

When seed tuber reach the third stage (normal sprouting) is regarded as ideal condition for planting. Normally seed potatoes are kept in cold stores at 2-4 degree C in bags (40-60kg capacity) or plastic/wooden crates. However, in hilly areas of NEH region shallow wooden/plastic crates are good for better shelf life.

At ambient temperatures in warm climates, the potatoes may reach the stage of senility after as little as three to five months. When choosing the seed source, the physiological stage of the seed will be in, is very important.

A tuber is in a dormant state if it does not sprout when exposed to conditions suited for sprouting, mainly a temperature above 4°C. The duration of the dormancy period is primarily a varietal character and is under hormonal control. Long tuber dormancy is very important for good keeping quality, although long dormancy alone does not ensure good keeping.

The physiological age can be influenced during storage and transport, especially by temperature. High and poor ventilation during transport can also lead to physiological death of the tubers i.e. the sprouting ability can be destroyed. Planting of physiologically old seed is a great risk for uniform stand and good yield.

(iii) Physical Defects

External damages may occur to tubers during harvesting including any form of subsequent handling. Serious damage can be caused also during transportation. Cuts, bruises, skin abrasions and sun & frost injury (to mention a few) will affect the storage. They also make the tubers vulnerable to bacterial/ fungal rotting. Planting of such seed material would lead to poor germination and plant growth.

Potato tubers are likely to have internal damages, due to poor storage conditions. This may further enhance the vulnerability to diseases (e.g. in case of hollow hearts, caused by fast irregular growth) or cause decay, resulting in rottage of tubers and loss of vigour (e.g. in the case of black hearts, which is caused by shortage of oxygen during storage). The above diseases / disorders cannot be seen from the surface of the tuber, hence seed lots have to be checked by randomly cutting few tubers.

(iv) Seed

Potato seed size is considered to be those tubers that are too small for consumption. Planting such small tubers, however, is risky because of the little reserve of nutrients and moisture that will help the sprouts and young plants through their initial stages. Moreover small tubers are also produced by degenerated seed stocks. Therefore fraction of very small tubers within a lot usually contains a higher percentage of virus infection than the average of the lot. The acceptable seed size for production of quality seed in hills is 25-125 g. However, it is not practical as well as economical to plant large size tubers because of their higher weight, they will increase the cost of cultivation. Due to presence of soil borne diseases like, bacterial wilt in most of the part of NE region cut tubers should not be used as planting material.

Planting

(i) Method

Tubers for planting should be planted in furrows made against the slope 50-60 cm distance between the rows. It is important to make the furrows and ridges against the slope or in zigzag manner to avoid soil erosion. Large size Tubers should be placed in furrows in upper side of the field slope and small tubers in downward side slopes at 20-25 cm distance between the tubers depending upon seed size. The tubers are placed deep if the soil is dry and shallow if the moisture is more. Seed tubers should be covered immediately with soil after planting making the ridges to height of 8-10 cm.

(ii) Planting Time

The recommended planting time has been depicted in the table next page.

Height above sea level	*Planting time*	*Harvesting time*	*Part of NEH states*
Plains (up to 500 m)	Sept-Oct.	Jan-Feb	Assam, Tripura, Manipur, Valley Dimapur district of Nagaland and lower parts of Arunachal Pradesh
Mid-hills	Feb-March	End June to end July	Slang district (basar area) of Arunachal Pradesh, Jowai sub division of Jaintia hills, some areas in Khasi hills
High hills (up to 2400 m)	Mar-April	July-August	Parts of Arunachal Pradesh, Upper Shillong, Mawphlang and Mairang area of Meghalaya, Mao and Ukrul area of Manipur, Blue Mountains in Mizoram
Very high hills (above 2400 m)	-do-	-do-	Tawang, Se la upper Subansiri, Siang and Kameng districts of Arunachal Pradesh

Fertilizer Requirements

The crop requires balanced dose of NPK for optimum production. Nitrogen contributes to the yield by enhancing the number of tubers, bulking rate, bulking period and delays the maturity. Phosphorus and Potassium are the second and third limiting nutrients in potato production. Phosphorus is involved in a wide range of plant processes like cell division, development of root system and enhanced crop maturity. Its application increases the number of medium size tubers per plant, which has a special significance for the seed crop. While application of K increases plant height, crop vigor and imparts resistance against drought, frost and diseases. It also increases the yield by increasing the number of large size tubers. FYM and chemical fertilizer can be applied simultaneously at planting. The recommended schedules of fertilizers are given below.

Fertilizer	*Source*	*Quantity to be applied (q/ha)*	
		Seed potato	
		At Planting	*At earthing up*
Nitrogen (N)	Ammonium sulphate	3.75 (75 kg N)	1.25 (25 kg N)
	or Calcium Ammonium Nitrate (CAN)	3.00 (75 kg N)	1.00 (25kg N)
Phosphorus (P_2O_5)	Single super phosphate	7.5 (120 kg P_2O_5)	Nil
Potash (K_2O)	Muriate of Potash	1.00 (60 kg K_2O)	Nil

The fertilizer should be mixed with soil to avoid direct contact of the seed tubers.

Intercultural Operations

The purpose of intercultural operations is, a) to remove weeds, b) to incorporate the supplemental N so that growth of the crop is boosted and full canopy cover is attained early, and c) to ensure adequate coverage of the emerging stolons. Weeding should be done as soon as the weeds appear. The final earthing up should be done when the plants are 10-15 cm high. While earthing up, the remaining dose of nitrogen (25kg/ha) should be applied. Herbicides like metribuzin @0.7 kg/ha (1.0 kg Sencor) as pre emergence or paraquat @ 0.5 kg/ha (2.5 litre Gramaxone) as post emergence treatment at about 5% emergence of potato plants for effective control of the weeds in potato crop.

Plant Protection

Important diseases and insect pests along with their control measures/management practices are described in the following table.

Common name and Pathogen	Symptoms	Control
Late blight (Phytophthorainfestans)	Late blight is the most dreaded fungal disease that causes heavy to summer and autumn potato crop in Meghalaya. The disease is characterized by water soaked lesions and cottony growth on lower surface of leaves; light brown elongated lesions on the stems and petioles, which encircle the stem. Infected tubers develop reddish brown, shallow to deep dry rot lesions. Flesh of affected tubers becomes "caramelized" with a sugary texture.	Cultivation of resistant varieties like K. Giriraj K. Megha, K. Jyoti K. Kanchan; use of disease free seeds preferably from disease free areas/ fields; prophylectic spray with 0.2% mancozeb as soon as the weather becomes congenial for disease development (tem. 10-20^{0}C; RH > 80%); subsequent sprays at 8 days interval; alternative use of metalaxyl based fungicides viz. Ridomil MZ 0.25%; haulms cutting at 75% disease severity.
Brown rot	Drooping of leaves leading to complete wilting and collapse of one or more branches. Vascular browning of tuber with white slimy bacterial ooze out on cut surface.	Use of healthy disease free seed, early planting of crop between second to third week of February and harvesting before second week of June, crop rotation with non-solanaceous crops like maize, garlic, cabbage, carrot, cauliflower, knol-khol and collection and burning of the diseased plants and infected tubers after harvest. Application of stable bleaching powder 12 kg/ha (6gms per

Common name and Pathogen	*Symptoms*	*Control*
		'Nur' of 5 sqm.) mixed with fertilizers in furrows at planting.
Black scurf (Rhizoctonia solani)	Sprout killing cankers in the underground parts and scurf (brown to black sclerotial bodies on tubers)	Tuber treatment with Organomercurial compound (OMC) (0.25-0.5%) for 30 minutes or with 3% boric acid for 30 minutes. Crop rotation with cereals; deep ploughing in fallow fields; use of disease free seed.
Potato leaf roll (PLR)	Primary symptoms are confined to top young leaves, which usually stand upright, roll and turn slightly pale/reddish/pink. Secondary symptoms develop when plants are grown from infected seed tubers. Such symptoms are rather prominent in older leaves. Infected plants have characteristics pale, dwarfed and upright appearance with rolling of lower leaves that turn yellow, brittle and are leathery in texture.	Plant virus free seed potatoes and multiply virus free seed only in aphid free areas/periods coupled with practices for ensuring sanitation, rouging of infected plants, if any and dehaulming the crop as soon as the vector aphid populations exceed critical limit of 20 aphids per 100 compound leaves. Build up of incoming aphid vectors can be controlled by application of suitable contact systemic insecticides.

Common name and Pathogen	Symptoms	Control
Potato mosaics	The infected plant shows light green or yellow patches on leaves that are result of yellowing of green tissues referred as " Mottling".The plants so affected exhibit a general stunting. Affected leaves show necrotic symptoms followed by death of entire leaf.	It can be controlled by strict sanitation in the field and also in the stores, right from the harvest to planting. Disinfect all field equipments by dipping in or washing with 3% trisodium phosphate or calcium hypochlorite (1% solution) and planting of disease free seed stocks from approved or reliable sources: use of resistant varieties, if available. Rouging of diseased plants along with their tubers and disposing them away from the field at least twice during the growing season. Dehaulming the crop before the aphids cross-critical level to enforce rigid control of the insect vector.
PESTS		
Aphids	They cause damage to the crop by sucking the sap from leaves and tender parts of the plants. Leaves curl downward, turn yellow, also transmit potato virus Y (PVY) PVA and leafroll (PLRV)	To control crop should be sprayed with foliar systemic insecticide like methyl Demeton 25EC or Dimethoate 30 EC at 0.03% concentration, soon after appearance of 1-2 aphids/100 compound leaves. Subsequent need based sprayings at 10-15 days interval up to dehaulming should be given. Cultural methods of control like removing the volunteer plants, weeds and other hosts

Common name and Pathogen	*Symptoms*	*Control*
		especially those with yellow flowers from within and around the seed potato plot are also helpful.
Potato Tubermoth	Newly hatched larvae feed on the foliage or stems by mining them with the result leaves and stem dried up. On tubers they make tunnel inside the tuber and feed on pulp. Due to the secondary infection by bacteria or fungi tuber generally rots.	In fields, spray of monocrotophos 40EC @1.5 liter in 1000 liters of water on the standing crop is recommended. Cultural methods like slightly deeper planting (10 cm) of healthy tubers, covering all the exposed tubers with soil at earthing up, collection of all the left over tubers at the harvest and removal of all volunteer potato plants are also recommended. In stores, the seed tubers should be covered with 2-2.5 cm thick layer of dried and chopped lantana leaves for protecting tubers from PTM attack. Use of granulosis virus (GV) is extremely helpful in reducing damage in store.
White grubs	These grubs are creamy white/off white caterpillars that damage the crop in two ways. Initially they feed on rootlets/roots and secondly, they do tuber damage after tuberization. These grubs make large, shallow and circular holes in tubers and render them unfit for marketing.	
Cutworm	Caterpillars of cutworms are dirty grey in colour and cause main damage to the crop by feeding on young shoots or under ground tubers at night. After tuberisation they feed on tubers by making deep circular	Spray the crop with any effective contact insecticide like Endosulphan or quinalphos solution of 0.1-0.2%. Soil application of Phorate 10G or Carbofuran 3G @2.5 –3.0kg a.i./ha in

Common name and Pathogen	*Symptoms*	*Control*
	holes affecting both tuber yield and its market value.	furrows at planting and also 50 days after planting in areas having enough soil moisture.
		Spray the crop and drench the foliage and ridges with Chloropyriphos (Dursban/Danuban) 20 EC @ 2.5l/ha in 1000-1200 liters of water.
Red ants	Workers of red ants feed on potato tubers by nibbling and making small but deep circular holes.	Spray the crop and drench the foliage and ridges with Chloropyriphos (Dursban/Danuban) 20 EC @ 2.5l/ha in 1000-1200 liters of water.

Dehaulming

Irrigation is stopped 10 days before haulms killing in the valley / low hills. The haulms should be killed with gramaxone @ 3 1/ha by 15-20th August in hills and 25th December to 15th January in valley depending upon the date of planting in the region. The crop should get duration of 80-90 days before dehaulming. Exposed tubers should be covered with soil immediately after haulms killing. The crop should be harvested 10-15 day after dehaulming. This will allow hardening of tuber skin. The practice of haulm cutting is very important for the seed crop.

Harvesting

The crop is harvested 10-15 days after haulms cutting for allowing hardening of skin of potato tubers to prevent disease infection through skin. Harvesting should be carried out on bright sunny days. All the cut/crack/bruised and damaged tubers should be removed. Late blight, pink rot and soft rot infected tubers or tubers infected with other rot causing microorganisms are removed and rejected.

Grading of Produce

The grading of produce is important because of marketing purpose and also for giving uniformity to the next emerging crop. The produce is graded into four grades i.e. large (>150 in hills), medium (80-125 g), seed size (40-80 g) and small (20-40 g). The chats or undersized (<20 g) should not be used for seed purpose.

Seed Treatments

Harvested tubers are washed first in water and then in 1% bleaching solution and again in water. Treat the seed tubers with 3% commercial grade of boric acid for 20 minutes by dip treatment. The boric acid solution prepared once can be used for 20 times. Tubers should be dried in shade.

Storage of Produce

Storage of potato tubers should be done only after proper drying. The potato can be stored in plastic/wooden crates or in gunny bags of 50 or 80 kg capacity. Provision for proper aeration should be made to avoid rottage. The bags are kept in the cold stores at 2-4 degree C in the plains while they are stacked in the country stores in the hilly region. It is important to turn the bags once in every two months in the country/cold stores. The store should be cleaned and sprayed with insecticide before storing the seed potatoes.

Important Steps for Quality Seed Potato Production

If the seed crop is to be grown, the following extra steps should be followed in addition to practices taken up for general potato production.

1. Select the fields, which are, either free from brown rot pathogen or have minimum inoculum in the field.
2. Apply bleaching powder @ 12 kg/ha at the time of land preparation to manage the inoculum of brown rot pathogen.
3. Use properly sprouted seed only. Do not cut the tubers.
4. Adopt plant protection measures against foliar diseases as per recommendations.
5. Apply a granular systemic insecticide such as Phorate 10 G @ 10 kg./ha at the time of of planting to prevent infestation of aphid vectors.
6. Spray the crop with Imidachloprid (Confidor) @ 400 g/ ha in 1000 litres water when the aphid count reaches 2 aphids/100 compound leaves. Repeat the spray at 10-15 days intervals based on the aphid population buildup in the crop and cut the haulms when its population reaches 20 aphids/100 compound leaves.
7. During crop season, the field should be inspected thrice i.e. 40-45 days, 60-65 days and 75 days to remove all off-type and plant showing mottling, mosaic, crinkle, necrosis and leaf rolling symptoms.
8. Cut the haulms at the ground level in June in hills and during January in plains.
9. Harvest the crop when the skin is firm.

10. Sort out the infected tubers and grade them according to recommended size.
11. Before storage the seed should be treated with 3% boric acid solution for 30 minutes.

Non-conventional System

Under this system nucleus seed production can be achieved through large-scale multiplication of disease free microtubers/microplants and true potato seed technology.

Potato Micro Propagation

(i) Initiation of Cultures

Cultures for micro propagation can be initiated from meristem tips or sprouts tips of pathogen-tested tubers or by *in-vitro* plantlets received through germplasm exchange. Alternately, culture may be initiated from auxiliary stem segments of pathogen-tested plants. In either of the above methods, to initiate culture, 2-3 explants are inoculated in a test tube (150 × 25 mm) containing about 10-15 ml solid propagation medium. The propagation medium contains basal MS salt and vitamins, GA_3 (0.1mg-^{l}), Calcium pentothenate (2mg-^{l}), NAA (0.01mg-^{l}), sucrose (30g-^{l}) and is solidified with 0.8% agar.

(ii) In-vitro propagule multiplication

Newly initiated culture is incubated at 25± 2^{0}C under 16 hours photoperiod of 150-w/mm^2 fluorescent lights. After 15-20 days in this condition apical/auxiliary buds of the explants grow into shoots and attain a length of 6 cm. These well-grown shoots now become ready for subculturing. In general, two methods are employed for in-vitro propagule multiplication.

(a) In-vitro multiplication of plantlets

Multiplication of in-vitro plantlets through nodal cuttings has tremendous potential. After every 20-25 days, in-vitro plantlets can be subcultured on fresh solid propagation medium. At every subculturing, a single *in vitro* plantlet yields on an average three nodal cuttings. Since micro-propagation can be carried out round the year under controlled conditions, 18 sub-culturing can be done in a year. Therefore, theoretically 3^{18} (387 million) plant-lets could be obtained in one year from single in-vitro plant. 25 + -Alternate method of mass micropropagation is liquid culture, when sufficient numbers of *in vitro* plantlets are obtained on solid medium in test tubes; these are cut into larger pieces (each with 1-2 nodes) and inoculated in liquid medium. Such 8-10 pieces are added in 250-ml conical flask having 20-ml liquid propagation medium. Composition of this medium is same as that of solid propagation medium except that it does not contain agar. Auxiliary buds from stem pieces grown into new plantlets and fills the flask within 15-20 days. Some laboratories use shaker speed 120-130 revolutions per minute to incubate liquid cultures. This ensures proper aeration and rapid growth of plantlets. The obtained plantlets, after hardening, can be directly transplanted in glass house/screen house and allowed to grow into normal size plants, producing normal tubers. Alternatively, these plantlets can also be utilized, after manipulating the media and environment, for microtuber production in laboratory condition.

(*b*) *Micro Tuber Production*

Microtubers are miniature tuber produced under sterile tuber-inducing in-vitro condition. Therefore, they are considered to be absolutely free of viruses and other diseases. These small, dormant tubers are particularly convenient for handling, storage and distribution. Unlike micropropagated plantlets, they do not need the time consuming hardening period in a green house, and can also be adapted easily to large scale planting in the field.

For microtuber induction the unspent liquid propagation medium is dispersed out of the container, and 40 ml of fresh tuber induction medium is poured under aseptic conditions.

The induction medium is based on MS basal supplemented with 44.38mm BA and 80 g^{1} sucrose, and culture are incubated under complete darkness at 20°C. Depending on the genotype used, microtuber begin to appear 1-2 weeks after incubation. Usually they develop 60-75 days of incubation depending on the genotype. In general, 15-20 microtubers with an average weight of 100-150 mg can be harvested from each conical flask/ magenta box. In view of their delicate structure and high value, it is preferred to plant microtubers in greenhouses.

(*c*) *Minituber Production*

Minitubers are produced from in vitro plantlets or microtubers after they have been planted at high density (from 100-200 plants per m^2) in beds in poly- or screen houses. The plants remain small and consequently small tubers are produced. The Minitubers thus produced, can be planted directly in the field and enter the seed production chain. These Minitubers are classified as pre-basic seed. The advantage however, is the fact that the tubers obtained have been less exposed to infections with diseases than the clones that have already been in the field for a number of generations. This may be of particular importance in NE region where high degeneration rates prevail.

True Potato Seed Technology

Besides the regular seed production systems potato production from true potato seed (TPS) has been possible in most of the potato growing areas of NE region. TPS can be used as a low input alternative for raising commercial crop especially in areas where availability of quality tuber seed at reasonable price is a major constraint. An eye catching advantage of this method is that the planting of one ha potatoes require only 100-150 g TPS instead of 20-25 t of tubers, which could result in a tremendous saving in cost of production. Added advantages are that the problems and high cost of transport and storage of potato tubers can be avoided by the use of TPS.

True seed produced is produced in the berries that are formed after the potatoes have flowered. Due to genetic recombination, each seed develops into a different plant with different characteristics, resulting in potato crops that are heterogeneous with regard to agronomical characters and tuber quality. Through selection several open pollinated and hybrid populations have been developed. TPS populations, TPS C-3,

HPS 1/13 and PT 92-27 are the few, being grown at farmers field.

Following two methods have been found successful for raising commercial potato crop using TPS derived planting materials.

I. Transplanting seedlings.

II. Planting seedling tubers.

Transplanting TPS seedlings is successful in areas where the winters are mild, irrigation is available, seedlings get established in shorter time and produce acceptable yields. Second method using seedling tubers as planting material is successful in all areas wherever potato is grown through seed tubers.

Production of potato crop from seedling tubers in the field

The recommended cultural and manurial practices of the region may be followed for potato production from seedling tubers.

(*i*) The different grades of seedling tubers produced in the previous crop season may be used as seed for raising the ware crop by adjusting inter and intra row spacing according to their size.

(*ii*) For mechanical cultivation, keep the inter-row distance at 60 cm and for manual cultivation at 45 to 50 cm. The plant to plant distance in a row may be kept according to the size of seedling tuber as below :

	Manual cultivation		*Mechanized cultivation*	
Size of seedling tuber (g)	*Spacing (cm)*	*Approx. Seed Rate (q/ha)*	*Spacing (cm)*	*Approx. seed rate (q/ha)*
20-40	45-50 × 20	30-33	60 × 20	25
10-20	45-50 × 15	20 –22	60 × 15	17
5-10	45-50 × 10	15 – 17	60 × 10	13

Follow the seed plot technique *viz.* timely planting, timely spray of pesticides and haulm cutting before the critical level of 20 aphids per 100 compound leaves is reached in the crop, if the produce is to be used for seed purposes. In traditional areas of seed production (*e.g.* Assam & Tripura plains), the produce can be retained as seed for another 3-4 years if seed plot technique is followed strictly.

Seed Standard and Certification of Seed Crop

It is a way of examining the stocks by visual means to identify and locate abnormal plants in the growing crop and tubers in the produce and removing them to maintain purity & disease freedom.

Potato is propagated sexually as well as asexually but the health of seed is the basic consideration and the system of seed production varies fundamentally from the grain crop.

Methods of Inspection For Certification

The crop should be fully-grown at the time of final inspec-tion and the germination should be uniform. Too early or too late inspection serves no purpose of seed certification. Final field inspection for certification is carried out provided the following conditions are fulfilled:

(*i*) The previous history of the field, crop operation, manuring and plant protection schedule are made available at the time of inspection.

(*ii*) Final roguing should be done before certification inspection. At the time of roguing all typical and diseased plants should be removed along with mother and newly born tubers.

Minimum four counts of 100 plants each are taken randomly on four spots in an area of one hectare. For additional each hectare or part thereof one sample of 100 plants is observed for all visible mosaics other diseases and off types. All relevant ob-servations of each count will be filled in the proforma. The overall appraisal of all the counts, thus inspected, will decide the rejection or selection of the seed crop in a particular field. If the crop meets the tolerance limit of Foundation-I or Foundation-II seed grade or certified seed then it is selected or passed for a particular category of seed.

If it does not meet the tolerance limit of particular grade then it is either down graded or rejected for ware use. After the dates are disclosed on the basis of actual aphid count for haulms cutting, certification authority is to en-sure that haulms at the ground level are removed on due dates and there is no re-growth of stamps at all. Finally the graded produce is inspected for tuber grades and surface infections.

Certification of Stocks

The seed certification methods are a way of ensuring the availability of true to type healthy seed, free from degenerative diseases and varietal mixture. The main objective of seed certification is to control the spread of fungal, bacterial, viral and nematode diseases and to maintain the quality of seed. Hence the general approach, which implies to potato, is given below:

1. Seed tubers should not be less than 30 g.
2. Seed stocks should be reasonably clean from black scurf and common scab.
3. Proportion of tuber with cuts bruises; cracks should not exceed 1% by count.
4. Seed should be free from insect damage, worms, late blight, dry or wet rots, charcoal rots, wart, black heart and freezing in-jury. The damage caused by above factors should not be more than 1% by count.

In North-eastern India only Assam State Seed Certification Agency is involved in certifying the seed of different crops.

In seed certification following aspects are taken into con-sideration for inspecting a crop:

(*i*) History of the plot about the previous crop, soil borne pathogen whether it is free from them or not.

(*ii*) Isolation of 15m between seed and ware crop has been maintained or not.

(*iii*) Roguing of diseased plants has been done or not and plant protection measures have been taken or not.

(*iv*) Variety & source of seed.

Other information on crop rotation, pre sprouting of seed stocks, cultural practices & timely planting, suitable combina-tion of seed size and spacing will give the idea about the varietal mixture (in hills only) tuber born diseases and the quantity of seed size tubers at harvest.

Seed Certification Standards

As per the national policy of seed multiplication, the mul-tiplication phases of seed have been grouped into three categories i.e. i) breeder's or basic seed which is produced scientifically under the supervision of expert Scientists, ii) foundation II & I. It is produced by State Agriculture/ Horticulture departments, NSC & SFCI and iii) certified stocks which is grown on contract with growers.

Viruses

The breeders seed is considered to be pure and disease free and no tolerance limit is fixed for it, while for foundation (FS I, & FS II) and certified seed stocks tolerance limit of viruses, off type, tuber born diseases and grades have been fixed by Govt. of India which are as follows :

Grade of seed	% of plants infected with crop/purity				
PSTV	*Off*	*Mild mosaic*	*Severe PLRV/PVT LST/Yellow*	*Total Viruses*	*Brown Rot*
FS-I	0.05	1.0	0.5	1.0	–
FS-II	0.05	2.0	0.75	2.0	3 pl/ha
Certified	0.10	3.0	1.0	3.0	-do-

Seed tubers diseases

Grade	% Incidence					
	*Common Scab**	*Black scurf*	*cut/ bruised*	*LB*	*Dry rot*	*Total disease*
FS-I	5.0	5.0	1.0	1.0	1.0	5.0*
FS-II	5.0	5.0	1.0	1.0	1.0	5.0
Certified	5.0	5.0	5.0	1.0	1.0	5.0

***Standard & limits approved 1982-83**

Seed standards

Hill Seed			
	Grade	Size	Corresponding weight
	Seed size	30-60 mm	25-150 g
	Large	Above 60 mm	Above 150 g
Plains Seed			
	Seed size	30-55 mm	25-125 g
	Large size	Above 55 mm	Above 125 g

NB : In a seed lot tubers not conforming to specific size of seed should not be more than 5%.

True potato seed crop

Mild mosaic	3%
Severe mosaic leaf roll and yellows	1%
PLSTV/PVT	0%
PSTVd	0%
Purity of seed	98%
Insert matter (max)	2%
Weeds seed	10/kg
Germination	80%
Moisture (max)	8%

Quality Control

The post harvest quality of stocks is checked by grow out tests. A sample of 2 tubers from each jute bag of 50kg is drawn and bulked as one lot based on the seed supplied to the seed-produc-ing agency. Thus these samples of each variety are planted lot wise separately and observations on disease incidence & varietal mixture are recorded and the quality is judged. Based on the percentage of viral diseases, the quality of seed is decided whether it is worth for Foundation-I or Foundation-II or certified grade. A few tubers are also cut to see the bacterial infection in the stock.

CHAPTER 16

TPS Technology : An Alternative Option to Seed Potatoes

Key-Features of Potato Multiplication Systems

S.N.	*Vegetative multiplication*	*Sexual multiplication*
1.	Based on whole tubers, tuber pieces, eyes, or plant cuttings.	Based on the botanical potato seeds, which are formed in the potato berry.
2.	Low multiplication rate, Approximately 1 : 10	• High multiplication rate, More than 1 : 750
3.	Easy transmission of several hundred diseases.	• Natural barrier to the transmission of all but five diseases.
4.	Avoidance of disease transmission can be limited, but is expensive and trouble some.	• Avoidance of disease transmission is easy and in expensive
5.	Planting rate varies from 1.5 to 3 MT/ha	• Planting rate varies from 60-250 gram/ha.
6.	High storage costs	• Low storage costs
7.	Rapid initial emergence	• Slow initial emergence
8.	Planting material that is perishable and can be stored for a few months.	• Planting material that can be stored for years without significant loss of viability.
9.	Planting window depends on the harvest time of the seed tubers	• Planting window does not depend upon the harvest time of the botanical potato seed.
10.	Traditionally accepted practice, and well understood by many in the agricultural community.	• A relative novel concept, which is not yet widely accepted by the agricultural community.

TPS Technology : An Alternative to Seed Tuber

The potato is a high input crop and is traditionally cultivated through seed tubers. About 40-50 % of the total cost of production is needed to procure seed tubers. The constraints of poor availability of high quality seed tubers in adequate quantity and at reasonable cost, inadequate availability of important inputs like fertilizers, pesticides etc., coupled with lack of knowledge of the improved technologies for potato production and inadequate facilities for storing seed potatoes in northeastern states are the major reasons responsible for poor potato productivity. It is in this context that TPS (True Potato Seed) technology assumes potential as an alternative method of potato production that is scientifically sound, technically feasible, economically viable, and ecofriendly to generate low cost and high quality planting material for enhancing the productivity, production and reducing the cost of potato cultivation. TPS, besides giving an almost disease free potato crop, is also cost effective. Being very small in size, TPS can be stored and transported easily with almost no cost involved in the same. In addition, only 50 gms TPS is required for sowing in about 350-375 sq. mt. area for producing seedling tubers enough for planting one hectare next year which in case of potato production through seed tubers would require 2-2.5 ton/ha. Hence, it significantly reduces the total cost of cultivation. In addition, the obtained produce of TPS crop could also be utilized as the quality seed stock for planting in subsequent season for mitigating the problem of poor availability of quality seed material in this region.

TPS is a sexually reproduced propagule in potato and results from the fertilization of ovules, which develop into tiny seeds inside the fruit called berry that are formed after the potatoes have flowered. Due to genetic combination, each seed develops into a different plant with different characteristics, resulting in potato crops that are heterogeneous in many aspects with regard to agronomical characters and tuber quality. Through selection, several open pollinated and hybrid varieties have been developed that to a certain extent to overcome these problems. The seed thus produced is called TPS or botanical seed to differentiate it from the conventional tuber seed. True seeds have the potential to develop into full-grown plants and produce tubers.

Naturally produced true potato seed results from self-pollination of flowers on potato plants. After potato plants reach maturity and flower, the flowers pollinate themselves when agents such as the wind or bumble bees cause the flowers to vibrate. They soon develop small, green fruit about the size of a cherry. Each fruit, or seedball, contains from 50 to 600 minute seeds. Simply crushing or grinding the seedballs extracts the seeds. Then the seeds are dried in the open air, under a fan, or even by placing a cheesecloth bag of seed in ordinary household clothes dryer. TPS is usually dormant when first extracted and thus must be stored for a few months, or treated with a growth stimulant such as gibberellic acid, before it will germinate.

Historical Background of TPS Technology

TPS is tantalizing because it can be grown easily in ordinary fields without resorting to the painstaking remote cultivation techniques required for seed tubers. However, in the past TPS would seldom produce good tubers, so the acquired wisdom of the centuries was that potatoes are best propagated by seed tubers.

After lying untapped for centuries, the potential value of TPS culture seems to have been realized independently at about the same time in China, New Zealand, the United States, and perhaps elsewhere, in the mid-1970s. U.S. Department of Agriculture (USDA) researchers in Washington State announced publicly in 1975 that, like vegetables, potatoes could be raised from TPS planted in fields. Besides developing the technology for growing crops from TPS, they tested TPS from several hundred varieties of self-pollinated potatoes to find a few varieties whose TPS will yield a crop of somewhat uniform type, size, and quality. With these findings the USDA researchers began promoting the idea that TPS could be especially valuable for developing countries.

Another major advance has been achieved in the Department of Horticulture at the University of Wisconsin, where Professor Stanley Peloquin and other researchers have developed complicated procedures for crossing divergent types of potatoes to produce hybrid seeds. This hybrid TPS produces plants that have highly desirable and carefully selected characteristics, such as high yield combined with resistance to diseases, pests, and environmental stresses, such as lack of adequate rain.

With new varieties and methods being developed for use in Third World countries, growers can harvest TPS from a field of potatoes, and later harvest the tubers for food. Importantly, subsistence farmers in developing countries can use all of the tubers for food, rather than saving quantities for next year's planting. TPS also stores well, even for several years. Unlike tubers, TPS does not rot, and loss during storage is small. TPS can be planted using the same methods and equipment typically used for tomatoes or other small-seeded vegetables. Usually, however, the seeds are allowed to germinate in pots or flats under cover, and the potato seedlings transplanted directly into the field.

Approaches in TPS Technology

The TPS can either be used directly for raising a commercial crop by first raising seedling in nursery beds and then transplanting them in the fields or by first producing seedling tubers in nursery beds/fields which are subsequently used for raising the commercial crop next year. However, due to the vulnerability of the seedlings at and soon after germination, sowing in plantbeds is preferred over direct sowing in the field, although it requires a tremendous amount of manual labour. Compared to these two methods, producing tuberlets in nurseries is the safest method. It, however still presents the disadvantages of the genetic variability and labour requirement, whereas some of the advantages, namely with respect to bulkiness and perishability in relation to transport and storage, are lost to a great extent. The first alternative of raising commercial crop by transplanting seedlings in the fields is found successful in Tripura and Assam only where winters are mild whereas the second method has been found successful in other potato growing areas of this region. However, the yield is higher when the crop is raised using seedling tubers as planting materials.

Although potentially attractive, TPS technique is still in a juvenile stage and cannot perfectly substitute the conventional methods used for seed potato production. However, it can co-exist as an alternative to seed tuber.

Advantages of TPS technology

TPS offers many advantages over seed tuber for various attributes of potato production. TPS can effectively overcome some of the problems associated with seed tubers and can be used easily by the resource poor farmers of the NEH region to produce healthy planting material in any quantity as and when required. TPS also offers many advantages to the framers to overcome weaknesses of clonally propagated tuber seeds.

Cost of production

Cost of planting material produced through TPS is approximately one tenth of the cost of quality tuber seed. In this technology only 100-150 g TPS is required costing about Rs 3000, in comparison to traditional method in which 3-4 tonnes seed tuber, costing about Rs35-40 thousands, for raising the crop in one ha.

TPS as source of healthy planting material

In TPS, except potato virus T (PVT) and potato spindle tuber viroid (PSTVd) no major pathogen is transmitted as they are filtered out during pollination and fertilization.

Saving seeds tubers for consumption

TPS saves nearly 18% of the total tuber retained as seed to be used for consumption. About 3-4 tonnes potato /ha used as seed tuber could be saved for human consumption. For a country like India, which produces more than 15 million tons of potatoes, almost 1.5 million tons are used as seed, which could have provided food to the millions.

Storage convenience

TPS with 3-5% seed moisture can be stored for many years under favorable conditions in dark with practically no loss in germinability at least up to 5 years.

Easy and inexpensive transportation

Since 1 g of TPS may contain 1500-2000 seeds; therefore only 100 g of TPS can replace 2-3 tonnes of seed tubers required for planting one-hectare land.

Potato cultivation in non-traditional areas

TPS can be used for potato cultivation in areas deemed unsuitable for producing quality seed tuber due to unfavorable agro-climate. Good quality and cheap seed tuber could be produced in areas where rate of degeneration is high and away from the main seed producing areas in the country.

Fitness of potato in different cropping systems

TPS can be used to fit potato in different cropping systems when tuber seed of correct physiological age is not available as and when required for planting the crop.

Environment friendly

Unlike the pathogens in clonally propagated crop are unable to affect the TPS crop due to in-built resistance (Multi-line effect) for diseases/pests. Consequently, lesser amount of pesticides is needed for spraying TPS crop. Thus TPS is not only cost effective but also environment friendly.

Disadvantages of TPS technology.

TPS also has some disadvantages, which have been the major bottlenecks in adoption of TPS technology.

- TPS produced crop takes about 15-20 days more for maturity compared to that from seed tubers.
- TPS is a labour intensive technology because of more farm operation viz. nursery raising, seedling transplanting and double earthing up etc.
- The crop has heterogeneity for various characters inherent in seedling populations.
- Potato seedlings are vulnerable to environmental stress and damage due to insect/pests and need more care/ labour input especially during the initial phases of growth and establishment in transplanted crop.

- Crop from TPS populations are less uniform in plant type/maturity, tuber shape, size and dry matter.
- The produce of TPS crop may not be acceptable to sophisticated consumers and the processing industries because of high preference for large size and uniformity of tuber characters. The present TPS hybrids produce medium size tubers and show some tuber flesh colour segregation.

Priority areas for TPS dissemination

TPS technology has a wider scope for its adoption in areas where quality tuber seed of a variety cannot be produced, yields are extremely low due to availability of poor quality seed, seed tuber storage and transportation are expensive, skilled labour is available and consumers do not have any preference for specific tuber characteristics. TPS technology is suitable in the NEH Region, where yields are extremely low (<10 t/ha) due to poor quality seed tubers. Moreover, as the region is unfit for breeder seed production, TPS could be a viable solution to mitigate the problems of non-availability of quality seed potato in this region. In addition, most of the potato farmers in this region are small and marginal. Hence, TPS could be appropriate and low-cost technology for them. The state like Meghalaya, Manipur,Nagaland and Arunachal Pradesh grow potato in two seasons. Hence, seedling tubers produced could be effectively utilized as the quality seed for the next season.

Agro-techniques for raising potato crop from TPS

TPS is the low cost source of healthy planting material and can be used in following three ways to grow the commercial potato crop.

- Direct seeding of TPS in the field.
- Transplanting TPS derived seedlings.
- Planting seedling tubers raised from TPS.

The direct seeding of TPS to grow commercial potato crop in the field has not been practically feasible due to the uneven and low percentage of germination, heavy weed problems, slow growth of the plants, etc.leading to poor yield. However, potato cultivation through transplanting seedlings and by planting first generation seedling tubers has been achieved successfully. TPS for planting should be viable, possesses desired vigour and is totally free from dormancy. Freshly harvested TPS has a dormancy period of 4-6 months depending on the genetic composition of the TPS population. The germination of TPS is well when minimum temperature is around 20±2° C. Therefore, time period of TPS nursery raising should be adjusted accordingly. An equally good germination could be obtained in slightly higher temperature ±5° C with the help of artificial shade during daytime.

Preparation of nursery bed for raising seedlings

1. Mark the nursery bed area keeping its breadth as 1m and length as per convenience. The beds may be separated by 80cm wide pathway.
2. The soil for nursery bed should be taken from a depth of 20-25 cm to avoid weeds and soil borne diseases. Mix the soil in 1:1 proportion with fully rotten FYM or compost. Add the basal dose of N, P and K at the rate of 13 gm urea, 75 gm SSP and 10gm MoP per square meter. In areas infected with brown rot, like Meghalaya, Arunachal Pradesh and Tripura, the level of nursery bed should be raised about 25 cms from the soil/ground.
3. Mix Thimet10G granules at the final nursery preparation stage to keep away ants and other insect pests.
4. Cover the top of nursery beds with half-inch thick layer of screened FYM.
5. Make proper arrangements to protect nursery beds from direct rains and sunlight.

   In areas where brown rot is endemic, the seedbeds should be raised by about 25-30 cm above the field level so that the seedlings do not root in the native infested soil.

Sowing of TPS in nursery beds

1. About 50 gms TPS and around 375 square meter nursery

bed area are required for producing enough seedling tubers for planting one hectare area in the next year. Soak the TPS in water for 24 hours, remove from water and again soak in 0.2% Diathane M-45 solution for 15 minutes. Then air-dry the seed in shade for 2 hours.

2. Mix the dried TPS with dry soil or sand or fine FYM. Spread the mixture evenly on nursery bed at a rate of 1-2 gms of TPS per meter square and cover it with half inch of the layer of FYM dust.
3. Irrigate the nursery bed using a sprayer ensuring that the soil is kept moist without any run off water.
4. Provide shade over the nursery bed from 10 AM to 5 PM if day temperature is above 30 C for next 15 days after sowing. Leaf or straw mulch or some other protection can be used to protect bed from direct sun light and rain. This hastens the germination by conserving moisture and reducing soil temperature.
5. Keep constant vigil of nursery bed to see requirements of seedlings. Irrigate the seedbeds as per the requirement to keep the bed moist.
6. Keep the nursery bed open during day and night when germination starts.
7. Foliar spray of 0.1% urea (one gram urea in 1 liter water) after 2-leaf stage is beneficial for quick growth and vigor of seedling. Seedlings are ready for transplanting in 30 days of sowing when it attains a height of 10-12 cm. Seedling must be transplanted before stolon initiation to avoid yield losses.

Transplanting of Seedlings

Transplanting of Seedlings in the field Under Irrigated Conditions :

1. Transplant seedling in field in afternoon only, to reduce damage due to water stress.
2. Broad cast 60 Kg N/ha (130 kg urea) and full-recommended dose of P (750 kg SSP) and K (100 kg MoP) at the time of final land preparation for seedling transplants.

3. If FYM application is done, first apply fertilizer mixture in furrows and then cover with FYM.
4. The furrows should be 4-5 inches deep at 40cms apart and prepare irrigation channels as practiced in the region.
5. Irrigate the furrows of two adjacent beds simultaneously just before transplanting. Take care to fill only half of the furrow with water leaving the top 8-10 cm of ridge dry.
6. Carefully remove the bare rooted seedlings from nursery beds and take them to the field in small baskets
7. Transplant the seedlings in two beds simultaneously (on both sides of irrigation channel) in north facing ridges by pushing seedling root in soft mud 10 cm apart and allow water to go in first furrow in 2nd bed. Then transplant seedling in first furrow of 2nd bed and allow the water to go in 2nd furrow of first bed. Thus, transplant seedlings in furrows of 1st and 2nd bed alternatively. The interplant spacing should not exceed 10cm for higher yield.
8. Spray 0.2% Dursban/Chlorpyriphos 20EC (25 ml in 10 liters of water) or Endosulfan 35EC (20 ml in 10 liters of water) on fourth day to protect seedlings from cutworms.
9. Irrigate the furrows every 3rd or 4th day till the seedlings get established. The frequency of irrigation can thereafter be reduced to once in 8-10 days as per crop need and according to the weather condition.
10. Do weeding after 25-30 days of transplanting depending upon the intensity of weeds and perform the earthing up operation and also apply ¼th dose of recommended Nitrogen i.e., 30 Kg N/ha (urea 65 kg/ha). Care should be taken to cover maximum nodes with soil leaving the top 5-6 leaves open in the air. The final dose of N (30Kg/ha) is applied at 2nd earthing up which is performed after 25-30days of first earthing up operation
11. Do all the subsequent operations including plant protection measures as per the standard potato cultivation practices of the region.

12. Dehaulm the crop 10-15 days prior to harvest and allow the skin to harden.
13. Harvest the crop after 110-120 days. The produce should be graded and small to medium sized tuber should be retained as the planting material for the subsequent season. The large size tuber, however, could be used for table purpose.

Transplanting of Seedlings in the Field under Rain fed conditions/Hills

1. Start transplanting of seedling when rains begin, approximately near 15th April. Therefore, the time of nursery raising should be adjusted according to the date of onset of rains in the region.
2. Open furrows at 40 cm apart and apply fertilizer @ 60Kg N, 120 Kg P2O5 and 60 Kg K2O per hectare. This is equivalent to 130 Kg Urea, 750 Kg SSP and 100 Kg MOP for one hectare.
3. Cover fertilizer with FYM @ 15 ton/ha followed by soil in such a manner that small ridges are formed.
4. Transplant seedlings on these ridges by making holes that are 10 cm apart. The transplanting should most preferably be done in afternoon.
5. Lightly irrigate the field if it does not rain after transplanting.
6. Spray 0.2% Dursban/Chlorpyriphos 20EC (25 ml in 10 liters of water) or Endosulfan 35EC (20 ml in 10 liters of water) on fourth day to protect seedlings from cutworms and other insects.
7. Filling of gaps created by mortality of seedlings in field should be done one week after the transplanting.
8. Weeding may be done 20-25 days after transplanting depending on intensity of weed and subsequently earthing-up operation should be performed with top dressing of Nitrogen @ 30 Kg or 65 Kg Urea/ha.
9. The second earthing up may be done after 25-30 days of first earthing up with third dose of N @ 30 Kg or 65Kg Urea/ha.

10. Do all the subsequent operations including plant protection measures as per the standard potato cultivation practices of the region.
11. Dehaulm the crop 10-15 days prior to harvest and allow the skin to harden.
12. Harvest the crop after 110-120 days. The produce should be graded and small to medium sized tuber should be retained as the planting material for the subsequent season. The large size tuber, however, could be used for table purpose.

Performance of TPS under Front-line Demonstrations (FLDs) conducted in North Eastern States

In order to evaluate on-farm performance of TPS technology, a total of 40 FLDs on TPS were conducted during the period of 2000-2002 across four states in northeastern region. State-wise, the numbers of FLDs were 30, 2, 6 and 2 in Meghalya, Nagaland, Assam and Tripura respectively. In all these FLDs, two TPS hybrids namely TPS C-3 and 92-PT-27 were utilized. TPS hybrids-wise, a total of 21 and 19 FLDs were conducted wherein TPS-C-3 and 92-PT-27 respectively were used.

Table 15.1 : Performance of TPS in different states in North Eastern states (NES)

S. No.	*States*	*Season*	*TPS Hybrids Used*	*Average Germination (%)*	*Average Yield (q/ha)*		*Average*
					Year 2000-02	*Year 2003-04*	
1.	Meghalya	Summer	TPS C-3	60	–	30	31.50
			92-PT-27	70	–	33	
		Autumn	TPS C-3	75	100	118	116.50
			92-PT-27	80	–	115	
			Average				**74.00**
2.	Nagaland	Winter	TPS C-3	60	32	–	121.00
			92-PT-27	80	–	210	
3.	Assam	Winter	TPS C-3	80	250	–	252.00
			92-PT-27	80	–	255	
Total/average yield across all the states					127.33	153.25	149.00

On-farm performance of TPS in NES revealed (Table 15.1) that during the reporting period, average germination percentage varied from 60 to 80 percent in the project villages. At nursery stage, incidence of late blight was noted in some cases particularly in the state of Meghalya. Similarly, mortality of transplanted seedling was found to the considerable extent more particularly in summer season in the states of Meghalya. The above table also disclosed that overall average yield of tuberlets was 149 q/ha with the break-up of 74 q/ha, 121 q/ha and 252.5 q/ha in Meghalaya, Nagaland and Assam respectively. Whereas in first year (2000) the average yields of tuberlets was 127 q/ha and it increased upto 153.25 q/ha in the next year i.e. 2001. Therefore, it could be inferred that the performance of TPS crop varied in different states. In the state of Meghalaya, where there are two main potato growing seasons viz. summer (Feb- June) and autumn (July-Nov), the performance varied greatly between the two seasons (Table 1). In Meghalya, the average yield of tuberlets on farmers' field ranged from about 32 q/ha (summer season) to 116 q/ha (autumn season). Similarly, average yield of tuberlets obtained from the two different TPS hybrids also varied but less considerably. The average yield of TPS C-3 and 92-PT-27 were found 30 q/ha and 33 q/ha in summer respectively and 118 q/ha and 115 q/ha in autumn respectively in the same state. Further, the performance of the TPS crop in Nagaland state was satisfactory with the yield 32 q/ha from TPS C-3 (year, 2000) and 210 q/ha from 92-PT-27 (year, 2001). However, the average yield of tuberlets in the states of Assam was still better figuring 250 q/ha for TPS C-3 and 255 for 92-PT-27. The beneficiary farmers planted the obtained tuberlets as the quality potato seed tuber in the subsequent crop season. The crop performed well with respect to growth, 1

Opportunities for TPS Technology

The commercial opportunities for botanical potato hybrid seed are substantial :

Potato, fourth largest crop in the world

First of all the potato is a major crop, which is grown in almost all countries of the world. The average area planted measures about 18 million hectares per year, so the demand for potato planting material is large. In most of these countries the supply of healthy potato planting material is either very expensive, or simply inadequate to meet the needs of the growers. In the northeast region almost 119 thousand ha area is under potato cultivation, which in turn requires a large demand for quality seed material.

Inexpensive to produce

Secondly the production cost of TPS is only a fraction of the amount that is needed to produce mini-tubers with a comparable health status. In addition to the much lower price structure, the TPS is also much cheaper to handle, store and transport.

Superior Health Status

The use of botanical potato hybrid seed also offers the potential to greatly improve the overall health status of the potato crop, and subsequently reduce the need for many chemicals. The use of TPS makes it commercially feasible to start with disease free planting material, which is not the case with seed tubers. Furthermore a field that is planted with botanical potato hybrid seed can offer more resistance against the spread of diseases through the field than a field planted with a clone. This is because the hybrid consists of a population of genotype.

Varietal change

The use of botanical potato hybrid seed allows for a much higher turnover rate of new potato varieties, and the much faster introduction of better traits both with and without the use of biotechnology. The turnover rate of traditional potato varieties is notoriously low. The estimated average life span of a potato variety is 60 years. The minimum time lag between the breeding of a clonal variety and commercial introduction is 12 years, with botanical potato hybrid seed that period be can greatly reduced.

Undeveloped potential

The study and use of botanical potato hybrid seed is a relative young branch of plant breeding and agronomy, which most likely means that the current status of the technology has not yet reached its peak. Various improvements can still be expected, especially in combination with biotechnology.

Constraints in TPS Production

As with any new product, the introduction of botanical potato hybrid seed will have to overcome several obstacles to develop as an alternative method of potato production that is scientifically sound, technically feasible, economically viable, and ecofriendly to generate low cost and high quality planting material for enhancing the productivity, production and reducing the cost of potato cultivation.

Certification of potato varieties

Potato varieties that are exclusively multiplied by means of cloning consist of only one genotype. Potato varieties that are multiplied by means of sexual potato seed consist of a population of genotypes. Since potato certification systems are based on clonal multiplication, it is currently not possible to certify the crop that results from a planting of botanical potato seeds. It will be necessary to set up a certification system that can accommodate the potato varieties that are produced grown from botanical potato hybrid seed. As long as such a system is not in place, it will remain a barrier for the wider use of botanical potato hybrid seed.

Consumers preference

In eastern Uttar Pradesh and Bihar, people prefer red skin tuber for consumption. They are ready to pay for it. In such a situation, it is difficult to push the technology with white skin TPS varieties.

Technology transfer

Successfully growing potatoes from botanical potato seed is very possible. It is a skill like any other that can be learned and transferred. As with the cultivation of any other crop, the highest yields are obtained by those who have the most experience, and professionals with experience in the horticultural aspects of the seedling year are still very scarce.

The majority of farmers that grow potatoes are unfamiliar with small seeded vegetables. For many of them the switch from a seed tuber to a real seed that is about 50,000 times smaller is often very difficult to make. The traditional growers of small seeded vegetables such as tomatoes, lettuce and cabbages have few problems in getting the seedlings started, but later in the seasons often lack the experience of the more seasoned potato grower. The best way of introducing botanical potato hybrid seed into an agricultural system is by engaging horticultural growers to produce seed tubers from botanical potato seed. These tubers can then be sold to the producers of ware potatoes, for whom virtually nothing will change except for a lower price and higher quality of the planting material.

Low risk bearing capacity

Risk bearing capacity of this technology specially up to seedling tuber production is very low in comparison to the traditional system of potato production. Therefore, once the farmers fail in raising the seedling in nursery beds, try to avoid this complicated system of seedlings tuber production.

Implications

Selection of the most suitable hybrid TPS varieties and subsequent availability of the required amount of TPS, either through importation or production within the country, would allow farmers to reap the advantages of this technology. The recommendations brought out by the Scientist define the future requirements for this technology. Policymakers will provide the needed policy incentives for the successful use of this technology for potato production and for the farming community to enjoy socioeconomic benefits together with enhanced sustainability for food production. Research should be on the right track to produce early bulking hybrid TPS families with built-in durable resistance to late blight and processing qualities. Partnership should be developed with the end-users for the further development of this technology and horizontal transfer to the clients—potato farmers.

The private sector and NGOs should also joined hands with the national program in making this technology successful in India. In near future TPS has to play a major role in providing the required quantity of quality planting material to contribute substantially in feeding the population in the 21st century.

CHAPTER 17

Package of Practices for Potato Cultivation in Meghalaya

Meghalaya- literally meaning the abode of clouds is one of the seven sister states of North Eastern Region of India. The state is stretched over a total area of 22,429 sq. kms, consisting predominantly of mountainous terrain with narrow valleys in between and strips of plain land in the south west. The state is home to about 23 lakh people, majority of them belonging to three ancient hill tribes of Khasi, Jaintias and the Garos. About 80 percent population of the state lives in rural areas and agriculture is their main stay.

As discussed earlier, rice, potato, maize, fiber crops, ginger, pulses and oilseeds are the major crops grown in the state. Potato is one of the major cash crops that forms an important part of the various crop sequences practiced in the state particularly in the high altitude areas of the state especially the central plateau of East and West Khasi Hills districts. Potato is grown as summer (March – July) as well as autumn crop (August- November) in the state. This crop is grown in about 20 thousand hectares of land, which is about 10 percent of the net area sown in the state. The state ranks second in terms of the total area under its cultivation and second largest producer of potato in the North Eastern Region

Despite the great importance in the state, the productivity of potato is 9.1 t/ha. The poor productivity figure in the state is attributed to an array of constraints existing in the state like rainfed conditions, hilly terrain, high incidence of diseases like late blight, brown rot, and insect pests like white grub, aphids, cutworms, Potato Tuber Moth (PTM) etc., poor availability of quality seed potato, continuing with the traditional methods of cultivation and poor knowledge about improved/recommended package of practices for potato cultivation.

This chapter, therefore, presents the recommeaded packages of practices for potato cultivation in this region under following sub-heads :

Recommended package of practices for Potato cultivation

Recommended cultivars

Kufri Jyoti, Kufri Megha, Kufri Giriraj, Kufri Girdhari and Kufri Hinalini.

Seed source

Seed should be purchased from a reliable source, preferably from a government agency. Purchase of certified seed should always be preferred and the seed stock should be replaced every 3-4 years to avoid use of degenerated seed potatoes.

Seed size and Seed rate

Seed tubers should be 30-50 g each in weight, having 3-4 healthy eye buds. The recommended seed rate is 30-35 q, however if small size seed tubers (30-40 g) are planted, the seed rate should range from 20-2.5 q/ha.

Planting time

The main crop i.e. summer crop should be planted during February to Mid March while the autumn crop should be planted in last week of August to first week of September.

Manuring

FYM : Well-decomposed farmyard manure (FYM) at the rate of 10-15 ton/ha should be applied in furrows opened for potato planting evenly.

Chemical fertilizers

The fertilizers should be applied at the rate of 60 kg N,

120 kg pP_2O_5 and 60 kg K_2O /ha at the planting time which in terms of urea, single super phosphate and murate of potash are 130 kg, 750 kg and 100 kg respectively. A second dose of N should be applied at the rate of 60 kg/ha (130 kg urea), as top dressing during the earthing up operation. For the local land unit used in Meghalaya called "*Nur*" which is usually 1m wide and 5 m in length (5sq m), application of 65 g urea, 375 g single super phosphate and 50 g murate of potash are recommended at planting time. At earthing up, a second dose of 65 g urea should be applied.

If FYM is applied at the rate of 10-15 ton/ha, N could be applied at a lower rate i.e., 100kg/ha (210 kg urea) in two split doses of 50 kg (105 Kg urea) each.

Bio-fertilizers

Nitrogen as well as phosphate fixing bio-fertilizers like Azotobacter and phosphate solubulizing bacteria can be used. Application of bio-fertilizers can reduce the use of chemical fertilizers.

Planting method

(*a*) Only well sprouted seed tubers should be used for planting. Do not plant hairy sprouted seed tubers.

(*b*) Seed tubers should be planted in furrows made against the slope to avoid soil erosion. The row to row spacing should be 50 cm and plant to plant spacing should be 20-25 cm. However, the plant to plant spacing between tubers could be adjusted as per the seed size. With smaller tubers the plant to plant distance of 15 cm should be maintained.

(*c*) Fertilizers in the right quantity should be thoroughly mixed to make a fertilizer mixture. After the furrows are prepared, apply the prepared fertilizer mixture in furrows then add well decomposed FYM and lastly place the tubers. Direct contact of tubers with fertilizer mixture should be avoided. Cover the seed tubers with soil immediately after planting and make ridges to a height of 8-10 cm.

(*d*) Use of cut tubers should be avoided to check spread of diseases from one tuber to another. If cutting of tubers for planting is unavoidable then the tubers should be cut longitudinally.

Interculture

Weeding : An important operation, which if timely carried out could lead to a good crop. It is done just before earthing up operation. Use of post emergence herbicide like paraquat @ 0.5 kg/ha (Gramaxone @ 2.5 liters /ha) in 800-1000 liters of water) at about 5% emergence of potato on weeds or pre-emergence herbicide like metribuzin @ 0.70 kg/ha (Sencor @1kg/ha within 3-5 days after planting could be used effectively to control the weeds in potato crop.

Earthing up operation

It is another important operation that should be done when crop attains a height of 10-15-cm. Use sufficient soil to cover the tubers. Proper earthing up has two benefits, firstly, it leads to production of more number of tubers and their development and minimize attack of potato tuber moth.

Rouging

During the growing season, the potato crop should be inspected to remove all off type and diseased plants showing molting, mosaic, crinkle, necrosis, wilting and leaf rolling symptoms if the crop is grown for seed purpose. Rouging should be thrice in potato crop.

The first inspection may be done when the plants attain 10-15 cm height or at 20 days after emergence for viral diseases and the second at flowering stage to check the varietal purity of the crop. The third inspection of the field should be done towards maturity to remove bolters. Bolters are the plants that germinate from tubers left in the fields in previous season. They can easily be identified as the plants that have thick and hardy stem and as they germinate later than the fresh planted tubers. The rouged out plants along with tubers should be burnt or buried deep in the soil to prevent further spread of diseases and they stay green when all other plants in the season approach maturity.

Insect pests management

Higher incidence of damage in crop by insect pests like cutworms, white grubs, defoliators and leaf-eating caterpillars is observed particularly during the dry and warmer seasons. PTM is another pest that affects the crop in field as well as tubers in stores.

Cutworm is an important pest of potato crop in Meghalaya. Caterpillars of cutworms are dirty grey in colour and cause main damage to the crop by feeding on young shoots or under ground tubers at night. In early stages of crop, caterpillars cut the stem of young plants near ground and feed on shoots and leaves. After tuberization, they feed by boring and nibbling into the tubers affecting both tuber yield and its market value. For management of heavy infestation of cutworms, drenching of foliage and ridges with Chloropyriphos (Dursban/Danuban) 20 EC@ 2.5 l/ha in1000- 1200 liters of water is recommended.

White grubs also cause significant yield losses in potato crop. These grubs are creamy white/off white caterpillars that damage the crop in two ways. Initially they feed on rootlets/ roots and secondly, they do tuber damage after tuberization. These grubs make large, shallow and circular holes in tubers and render them unfit for marketing. The population of white grubs could be reduced by spraying the beetles host trees/ shrubs with effective contact insecticide like Endosulfan or quinalphos solution of 0.1 – 0.2 %. Soil application of Phorate 10G or Carbofuran 3G @ 2.5-3.0 Kg a.i./ha in furrows at planting and also 50 days after planting in areas having enough soil moisture, is also recommended for management of white grub population.

For control of above mentioned soil pests, 2-3 deep ploughing of the field in autumn season should be done to expose pests for natural mortality and predation.

For control of the defoliating pests like leaf eating caterpillars in potato crop, spray of Endosulfan @ 1.5 liters/ha or Carbaryl @ 2.0 Kg/ha in 1000-1200 liters water is recommended.

Potato Tuber moth (PTM) is another important pest in the state. In fields, the PTM larvae penetrate leaves and feed within leaf veins or stems of plants. In storage, the larvae cause damage by depositing eggs on the eye buds. The larvae bore into the pulp of the stored potatoes through minute holes that become blackish within 3-4 days due to accumulation of faecal matter near entrance holes. The potatoes thus become unfit for use as seed or for human consumption. The infested tubers are further exposed to bacterial infection that leads to rotting.

For management of PTM in fields, spray of monocrotophos 40EC @ 1.5 liter in 1000 liters of water on the standing crop is recommended. Cultural methods like slightly deeper planting (10 cm) of healthy tubers, covering all the exposed tubers with soil at earthing up, collection of all the left over tubers at the harvest and removal of all volunteer potato plants are also recommended for control of PTM damage in fields. In country stores, the seed tubers should be covered with 2-2.5 cm thick layer of dried and chopped lantana leaves for protecting tubers from PTM attack. Use of granulosis virus (GV) is extremely helpful in reducing damage in store.

Aphids are small pests that cause damage to the potato crop by sucking the sap from leaves and tender parts of the plants. The direct damage caused by aphids is only marginal but they cause great losses by transmitting viral diseases and thereby limit the production of disease free seed potato production. Therefore, in crops particularly meant for seed purpose the infestation should be controlled. The aphid population in Meghalaya is maximum in summer crop i.e., the main crop, where it crosses the critical level of 20 aphids/ 100 compound leaves in the second week of May and remained high till maturity of the crop. In autumn potato crop it crosses the critical level in the first week of October.

In order to control the population of aphids, the crop should be sprayed with foliar systemic insecticide like methyl Demeton 25 EC or Dimethoate 30 EC at 0.03% concentration, soon after appearance of 1-2 aphids/ 100 compound leaves. This happens usually by the first week of May. Subsequent need based sprayings at 10-15 days interval up to dehaulming should be given. The spray is must for the potato crop meant for seed purpose. Cultural methods of control like removing the volunteer plants, weeds and other hosts especially those with yellow flowers from within and around the seed potato plot are also helpful.

Management of Diseases

Late blight is the most important fungal disease that causes heavy damage to summer and autumn potato crop in Meghalaya. The disease is regular and appears usually during second fortnight of May in the summer season and the entire foliage is damaged in 3-4 weeks. In autumn crop the attack is with a lower severity. The late blight disease affects the leaves, stems, and tubers of the plant. On the leaves small pale green spots appear that enlarge into large water soaked lesions. A white mildew (cottony growth) ring forms around the dead areas on lower surface of the leaves. In dry weather these lesions turn necrotic brown. On stems, light brown elongated lesions are formed which may encircle the stem. Infected tubers develop reddish brown, shallow to deep dry rot lesions. Flesh of affected tubers becomes “caramelized” with a sugary texture. The infected tubers are further attacked by other diseases. The infected tuber when pressed becomes hard. In case the tuber if pressed is soft, the disease is soft rot.

For management of late blight, cultivation of blight resistant cultivars like Kufri Giriraj, Kufri Megha, and Kufri Jyoti are recommended. Thorough sorting of diseased tubers, practicing haulm cutting when disease kills about 75 percentage of the foliage and removing of volunteer plants also help in checking the spread of this disease. Prophylactic spray of contact fungicides like 0.2% mancozeb (2 g of mancozeb per litre of water) is recommended as soon as the weather becomes congenial for its appearance (usually after 15 of May). A sticker like Triton AE @ 0.1 % (1 ml in one liter of water) must be mixed along with the fungicide. Subsequent sprays should be done depending on the weather conditions at 8 days interval till crop maturity. Alternatively, systemic fungicide like metalaxyl (Ridomil MZ 0.25% or 2.5 g of Ridomil in one liter of water) should be sprayed alternating it with mancozeb. Spray of mancozeb should be applied only after 15 days after

Ridomil application. However, spray of Ridomil should be restricted to a maximum of two sprays.

Early blight is another fungal disease of lesser significance as compared to late blight in the state. It manifests itself by appearance of small, circular to oval, brown spot on lower and older leaves. Mature lesions become large concentric rings looking like "bulls eye" and foliage looks dry and papery in contrast to the late blight spots that appear damp. To prevent the incidence of early blight, the crop must be given balanced doses of fertilizers, especially nitrogen and cultivation of solanaceous crops, being collateral host in nearby potato fields must be avoided. The chemical control of this disease is same as that for late bight.

Bacterial wilt or Brown rot is another major disease affecting potato crop in Meghalaya. It is the major disease that affect potato crop in field as well as during post- harvest handling, transit and storage in the state. It appears in both the summer as well as the autumn potato crop. It causes losses in two ways- firstly by premature wilting and death of plants thereby losses in yield and secondly by rotting of tubers on transit or storage. The disease symptoms in the standing crop include slight wilting of top leaves during clear sunny day, followed by sudden total wilting and collapse of one or more branches. In advanced stages, if the base of the stem of affected plant is cut transversely and squeezed, the bacterial mass is seen to ooze out as a dull white slimy mass on cut surface. In infected tubers, the vascular tissue in the cross section look like a water soaked circle followed by distinct brown discolouration and decay localized in the vascular ring. When the cut tuber is left for 3-5 minutes or a slight pressure is applied, a dirty white slime containing bacterial mass may appear on the vascular ring. In advanced stages, the bacterial mass may appear in the form of ooze coming out of the tuber eyes.

Absolute control of this disease is difficult however, following cultural practices like use of healthy disease free seed, early planting of crop between second to third week of February and harvesting before second week of June, crop rotation with non-solanaceous crops like maize, garlic, cabbage, carrot, cauliflower, knolkhol and collection and burning of the diseased plants and infected tubers after harvest are useful for its effective management. Application of stable bleaching powder @ 12 kg/ha mixed with fertilizers in furrows at land preparation stage or at planting also helps in reducing the bacterial wilt incidence by 80 percent.

Viral Diseases, also called as degeneration diseases are the diseases that cause deterioration of the seed stocks if the same stock is utilized continuously for number of years. The crops taken from such deteriorated seed stock is poor in vigour and the tubers are smaller causing yield losses.

Mosaics are the most common viral diseases that cause significant crop losses to the potato crop. The infected plant show light green or yellow patches on leaves that are result of yellowing of green tissues referred to as "Mottling". The plants so affected exhibit a general stunting. Affected leaves show necrotic symptoms followed by death of the entire leaf. Majority of the potato viruses except Potato Leaf Roll Virus cause mosaic mottling. Leaf roll is another serious viral disease of potato where the infected plants produce only a few, small to medium sized tubers. All Indian potato varieties are susceptible to this virus. Potato Leaf Roll Virus (PLRV) invokes two types of symptoms in plants depending on the age of infection, viz. Primary or Secondary. The primary infection develops within the crop season and usually confine to top young leaves that usually stand upright, roll and turn slightly pale in some cultivars. A reddish or pink colour appears in top leaves starting at margins, sometimes accompanied with slight rolling of leaves. Secondary infection develops when the plants are grown from the infected tubers and the symptoms are prominent in older leaves. Infected plants have characteristic pale, dwarfed and upright appearance with rolling of lower leaves that turn yellow, brittle and are leathery in texture. A reddish or purple discoloration develops on the margins and underside of the leaves in some cultivars.

Management of these viral diseases mainly lies in planting of healthy disease free seed stock purchased from approved or reliable source. The strict sanitation in field and in stores right from harvest to planting, helps prevent spread of viruses. The rouging out diseased plants along with their tubers and disposing them away from field at least twice during the growing season, destroying potato volunteers and weeds hosts before planting and practicing dehaulming before the aphids population cross-critical level of 20 aphids/ 100 compound leaves minimizes spread of viruses. In addition the build up of aphid vectors should be controlled by application of suitable insecticide.

Haulm Cutting

Haulms should be cut at the ground level in the beginning of June. At this time, the crop also starts maturing. The cut haulms should not be left in the field. Ensure that there is no growth of stumps after dehaulming as tender and succulent leaves are even more attractive to aphids. In addition, haulm cutting should also be performed if foliage is killed by late blight. This prevents the spread of blight percolating down in tubers and infecting them. The practice of haulm cutting is very important for the seed crop.

Harvesting

Harvesting should be done after 15 days of haulm cutting for allowing hardening of skin of potato tubers to prevent disease infection through skin. However, in Meghalaya, the potato crop is killed before attaining this stage due to late blight. Harvesting should be carried out on bright sunny days. All the damaged and rotted tubers should be sorted out and marketable tubers should be graded in different sizes as mentioned below :

Large	>150 g,
Medium	80 – 150 g,
Seed size	30 – 80 g,
Small	10 – 40 g.

Tubers below 10 g should never be kept for seed purpose as it may be infected with some disease that caused its small size.

Seed treatment

Produce that is to be kept as seed purpose should be first thoroughly washed with water, then treated with 1 percent bleaching powder solution and again washed with water, this practice is done to clean the tubers, the second step involves treatment with 3 % commercial grade boric acid for 20 - 30 minutes to control surface born diseases. Once prepared, the solution should be used 20 times for maintaining its effectivity and economics. Dry the treated seed in shade for two days. Drying of tubers in direct sun should be avoided. Treated tubers, being poisonous should not be used for table purposes.

Storage

Storage of potato tubers should be done only after proper drying. The storage should be done in wooden crates or plastic trays or in bamboo baskets in cool and dry place. Provision for proper aeration should be made to avoid rottage. Seed potato should not be stored in gunny bags as it encourages development of etiolated and lanky sprouts. Sorting of rotted tubers in store should be done as and when required. If these wooden crates/ plastic crates etc. are not available, spreading of potato tubers on dry surface in 10-15 cm layers is also recommended. However, storing potato seed in heaps should be avoided at any cost. Label the potato as table and seed while in storage. The store should be cleaned and sprayed with insecticide before storing the seed potatoes.

CHAPTER 18

Management of Major Potato Diseases

Management of Late Blight

It is one of the most important fungal diseases that cause heavy damage to both summer and autumn potato crops in Meghalaya. The disease is regular and severe in this state as weather conditions are most conducive for early appearance of the disease in the main cropping season and can prevail for a longer duration right from May to September/October. Usually, it appears during mid of May and the foliage is damaged within 3-4 weeks. The entire crop is killed before it achieves full maturity leading to heavy crop losses ranging from 50-95 percent in the summer crop depending upon planting time and variety. In addition, it causes yield reduction up to 50 percent depending upon the stage of the crop at which it is attacked. In autumn, the crop is also attacked by this disease however, the severity is low. The disease is caused by *Phytophthora infestans* (Mont.) de Bary.

Disease symptoms

In infected plants

- The disease affects all the plant parts *i.e.*, leaves, stems, and tubers.
- Small pale green spots appear on the leaves that enlarge into large water soaked lesions. A white mildew (cottony growth) ring forms around the dead areas on lower surface of the leaves. In dry weather these lesions turn necrotic brown.
- On stems, light brown elongated lesions are formed which may encircle the stem.

In infected tubers

- Tubers develop reddish brown, shallow to deep dry rot lesions. The flesh of affected tubers becomes "caramalized" with a sugary texture. Frequently metallic tinge develops on margin of affected tissue.

Management of late blight

(i) Cultural methods

These methods mainly aim at eliminating or reducing the initial inoculums load in the seed tubers and other sources in order to check the spread of the disease. They include

- Late blight resistant cutivars like Kufri Giriraj, Kufri Megha and Kufri Jyoti should be cultivated.
- Thorough sorting of diseased tubers should be done before storage and again before planting to reduce the chances of spread of the disease. The infected tubers should be collected and buried deep in soil.
- Ridges should be made high enough to cover all the daughter tubers thoroughly with soil to prevent growing out of fungus from mother tuber and also infection in daughter tubers.
- When 75 percent of the crop foliage is killed by the isease, the haulms must be cut, removed from field and buried deep. This practice however should be thoroughly done in seed crop to prevent further spread of disease.
- Ground keepers or volunteer plants harbor the pathogen and therefore should be removed immediately after emergence.

(ii) Chemical control

- As soon as the weather conditions become congenial for the disease appearance (usually after 15th of May), the crop should be given a prophylactic spray with 0.2% Mancozeb or Diathane M-45 or Indofil, which are contact fungicides (2.0 kg fungicide per ha. or 2 gms of fungicide per liter of water).

- A second spray of Metalaxyl based systemic fungicide viz., Ridomil MZ or Matco (0.25% or 2.5 Kg of fungicide per ha. or 2.5 gms of fungicide per liter of water) should be done 8-10 days after the 1st spray of Mancozeb. If the problem of blight still persists, 1-2 sprays of contact fungicide Mancozeb at 10-15 days interval is recommended (according to the requirements). The spray of Mancozeb should be done only 15 days after the Metalaxyl application as the effect of Ridomil/ Matco lasts for 2 weeks.
- While spraying Mancozeb, it should be ensured that the lower surface of foliage is thoroughly drenched with the fungicide solution. If spraying is done during rainy days, stickers like Triton AE (0.1%) or 1 liter/ha (10 ml in 10 liters of water should be mixed along with fungicide to prevent washing.

Management of Bacterial Wilt/Brown Rot in Potato

It is one of the major bacterial diseases affecting potato crop in field as well as during post- harvest handling, transit and storage in the states of north eastern region particularly in Meghalaya. This disease appears on summer as well as autumn/winter potato crop. Its incidence has been observed to vary from 10-12 percent in the summer and autumn/winter potato crop. The temperature of 25° to 35°C and moisture content above 50 percent favour faster development of this disease. The tuber rot is proportionately higher in relation to wilt in the state. *Ralstonia (Pseudomonas) solanacearum* Smith is the causal organism of this disease.

Disease Symptoms

(i) In Infected Plants

- The earliest symptoms are slight wilting of top leaves during clear sunny day. It is soon followed by sudden total wilting and collapse of one or more branches.
- In advanced stages, if the base of the stem of the affected plant is cut transversely and squeezed, the bacterial mass is seen to ooze out as a dull white slimy mass on cut surface.

(ii) In Infected Tubers

- Two types of symptoms are produced in infected tubers *viz.,* vascular rot and pitted lesions.
- In vascular rot, the cross section of infected tubers vascular tissue look like a water soaked circle followed by distinct brown discolouration and decay localized in the vascular ring. When the cut tuber is left for 3-5 minutes or a slight pressure is applied, a dirty white slime containing bacterial mass may appear on the vascular ring. In advanced stages, the bacterial mass may appear in form of ooze coming out of the tuber eyes.
- The pitted lesions on tubers are produced due to infection through lenticels (skin pores). Initially water soaked lesions develop at the site of the infected lenticels. The lesions enlarge to 0.5 to 1 cm in diameter and form 5 mm deep pits. The tubers skin remains intact but become shriveled.

A Simple Test

The presence of bacterial wilt infection can also be diagnosed by a simple test. Cut 3-6 cm long piece from the base of the wilted plant and suspend it with a thread in a glass tumbler containing clean water. Keep the tumbler undisturbed for 2-5 minutes. Bacterial wilt infected stem show whitish thread like mass of bacteria oozing out in water and the water at the bottom of the tumbler turns milky.

Management of the disease

(i) Cultural Methods

Absolute control of this disease is difficult; the following practices are however, useful for its effective management :

- Infected seed tubers and soil are the primary sources of infection carrying infection from a season to next. It is therefore, important that healthy disease free seed should be purchased and used as seed.
- Planting of cut tubers should be checked to avoid spread of the disease to healthy tubers.

- Early planting of the crop (between second to third week of February in hills) and harvesting before second week of June also considerably reduces the incidence of wilt.
- Crop rotation with non-solanaceous crops like maize, garlic, cabbage, carrot, cauliflower, knolkhol is also recommended.
- As the pathogen survive in most remnants and weeds during off-season, therefore soil should be ploughed twice at 20-30 days interval during winters
- Soil burning is also recommended in sick and infected fields.
- Collection and burning of the diseased plants and infected tuber after harvest is recommended.

(ii) Chemical Control

Apply stable bleaching powder @ 12 kg/ha mixed with fertilizers at the time of planting.

CHAPTER 19

Management of Major Insect Pest of Potato

The problem of pest cause significant damage to potato in the entire states in north eastern region of India including Meghalaya. The insects causing major damage to the potato crop are mainly cut worms, white grub, potato tuber moth, leaf eating caterpillars, aphids and red ants.

The following paragraphs, therefore, highlights some of the major insect pests of potato along with their symptoms and management practices.

Soil Pests

*Cut worms (***Agrotis ipsilon** *and* **A. segetum***)*

Cutworm is an important pest of potato crop in these states. Caterpillars of the cutworms mainly cause damage to the potato crop. The caterpillars are dirty gray in colour and feed at night on young shoots or under ground tubers. In early stages of crop, caterpillars cut the stem of young plants near ground and feed on shoots and leaves. After tuberization, they feed by boring and nibbling into the tubers affecting both tuber yield and its market value. The damage of cutworms is higher when days are warmer and drier after planting.

Control measures

(i) ***Cultural control***

- Deep ploughing in autumn season in hills in summer season in plains in these states is recommended for minimizing the cutworm population in immature stages.

(ii) Chemical control

- Drenching of foliage and ridges with Chloropyriphos (Dursban/Danusban) 20 EC@ 2.5 l/ha in 800-1000 litre of water after noticing cutworm attack on the crop.

White grub (Brahmina (Lachnosterna) cariaceae*)*

The larvae of the Scarabaeoid beetles, popularly called as Cockchafer or Chafer beetle are commonly called white grubs. They cause significant yield losses in the potato crop in this region. These grubs are creamy white/off white caterpillars that damage the crop in two ways. Initially they feed on rootlets/roots/FYM and second and third stage grubs feed on tubers. These grubs make large, shallow and circular holes in tubers and render them unfit for marketing.

Control measures

(i) Cultural control

- Deep ploughing of fields during autumn season in hills areas and summer in plains of the region is recommended for minimizing their population as it exposes the immature stage for predation by birds.
- Application of well rotten farmyard manure/compost also helps in managing the population of white grubs.
- Electric or petromax light traps should be used for mass trapping of beetle (adults) of white grub.
- Flooding of the potato fields, whenever possible, for 7-10 days in plain areas of the region could be practiced for effective control of white grub.

(ii) Chemical control

- The population of White grub could be reduced by spraying the beetles host trees/shrubs with effective contact insecticide like Chloropyriphos 20 EC @ 2.5 lit/ha in 800-1000 litres water, Endosulfan 35 EC or Carbaryl 50 WP @ 2.5Kg/ha in 1000-1200 litres of water at 0.1-0.2% concentration.
- Soil application of Phorate 10G or Carbofuran 3G @ 2.5-3.0 Kg a.i./ha in furrows in planting or near plant base

at earthing up time could be done in areas having enough soil moisture, also reduces the grub population

Defoliators

Leaf eating caterpillar (*Spodoptera litura* Fab., *Heliothis armigera* Hb. And *Plusia sp.*)

Leaf eating caterpillar is an important pest in all the states of North Eastern Region of the country including Meghalaya.

Epilachna beetle (*Henosepilachna viginitioctopunctata* Fab. ar d *Epilachna ocellata* Redtenbachor)

It ıs another important defoliating pest in these states. However its damage to potato crop in the Meghalaya state is not much.

Control measures

(i) Cultural control

- Hand picking of various stages of defoliators and their destruction should be taken up when their population is manageable.
- Crop rotations with non-host crops should be taken up.

(ii) Chemical control

- Spraying of the crop with Endosulfan @ 1.5 litres/ha or with Carbaryl @ 2.5 Kg/ha in 1000-1200 litres water is recommended for control of these defoliating pests.

Storage Pest

Potato Tuber moth (PTM)

The moths are small, pinkish in colour with grayish brown wings having small dark specks. The larvae moth causes damage to potato both in fields and the stores. In fields, the PTM larvae penetrate leaves and feed within leaf veins as leaf miner or stems of plants. In storage, the larvae causes damage by depositing eggs on the eye buds. The larvae bore into the pulp of the stored potatoes through minute holes and make tunnels inside. The entrance point become blackish within 3-4 days due to accumulation of faecal matter near entrance holes. The potatoes thus becomes unfit for use as seed or for human consumption. The infested tubers are further exposed to bacterial infection that leads to rotting. The extent of damage to the stored potato is found highest in the state of Meghalaya among all the other states in the North Eastern Region of the country.

Control of PTM

(i) Control of PTM in fields

(a) Cultural control

- Healthy tubers should be planted slightly deeper (10cms) than the conventional planting depth of 6 cms. This should be followed by proper earthing up in time i.e., covering all the exposed tubers with soil.
- Collection of all the left over tubers at the harvest as well as removal of all volunteer potato plants as PTM can harbour on them.

(a) Chemical Methods : Spray 30 days old crop with monocrotophos 40EC@ 1.5 litre in 1000 litres of water or with endosulfan 35EC @ 1.5litre in 800- 1000 litres of water. The spraying may be repeated two weeks before harvesting, if needed.

(ii) Control of PTM in stores

- Only Healthy tubers should be kept in stores. Use 2-3cm thick layer of chopped and dries Lantana or Eucalyptus leaves below and top of the potato heaps in the store to check PTM infestation.
- Use of sex pheromone traps @ 4 traps/100m^2 store for mass trapping of male tuber moths.
- Use Granulosis Virus (GV) formulations (to prepare 1 kg formulations, mix 1 kg talc powder in one litre of water and 20 GV infected PTM larvae and 5 ml triton in shade to make the formulation in to dust form). 5 kg of GV formulated dust is sufficient for one ton of potato.

Sap sucking pests

Aphids (Myzus persicae Sulzer and Aphis gossypii Glover)

These are small pests that by sucking the sap from leaves and tender parts cause damage to the potato crop. The direct damage caused by aphids is only marginal but they cause great losses by transmitting viral diseases and thereby limit the production of disease free potato seed production. Therefore, in crops particularly meant for seed purpose their infestation should be controlled. It is an important vector that causes significant damage to potato crop in the state.

Control of aphids

(i) Cultural Methods

- All the weeds, volunteer potatoes and other hosts susceptible to viruses and for aphids especially those with yellow flowers should be removed from within and around the seed potato plot.
- Rogue out the diseased plants showing virus symptoms from the seed plot three times in hills during the entire crop length first at 40-45 days after planting, second at flowering time (70-75 days after planting) and third when the crop is about to mature (100-105 days after planting).
- The seed crop should be dehaulmed at the right time after required maturity of crop but before the population of aphids reaches the critical level i.e., 20 aphids/100 compound leaves. Care must be taken to prevent any regrowth of stumps before harvests.

(ii) Chemical method

In plain areas of the statse, under irrigated conditions, application of granular systemic insecticides like phorate 10G @ 1.0- 1.5 kg a.i./ha has been found effective for 45-60 days provided there is sufficient soil moisture, therefore these should be applied at planting or earthing time depending on the appearance of the aphids in the area. This should be followed by 1-2 need based foliar sprays of suitable systemic insecticide such as methyl demeton 25 EC or dimethoate (Rogor) 30 EC @ 0.03 % concentrate in 1000-1250 litres of water/ha.

CHAPTER 20

Assessment and Refinement of Traditional Method of Potato Planting in Meghalaya

On-station as well as on-farm experiments were conducted to assess the farmers' traditional method of potato planting in comparison to the recommended and refined method. Three treatments *viz.*,**T1** : **Farmers' Method** (Raised Bed: L = 5 mts, W = 1.20 mts; P-P = 20cms; R-R = 40 cms and seed rate/ ha = 50 q/ha (1,25,000); **T2** : **Refined method** (Raised Bed: L = 5 mts, W = 1.20 mts; P-P = 20 cms; R-R = 60 cms and seed rate/ha = 33.30 q/ha (83,300) and **T3** : **Recommended Method** (Plot size: L = 3mt, W = 3.6 mt ; P-P = 20 cms and R-R = 60 cms and seed rate/ha = 33.30 q/ha (83,300) in seven replicates of each treatment planted in RBD were experimented on research station as well as farmers' field to find the comparative performance of each method.

(*i*) Findings

(a) Comparative performance of yield related attributes under three different methods of potato planting showed that average germination percentage was significantly more in the refined (95%) as well as the scientific method (90%) as compared to the traditional method (86%). Further, the average plant height was found more in the traditional method which is undesirable yield attribute. Similarly, with respect to the average number of branches and average numbers of the physiologically active compound leaves, no significant difference in all the three methods were noted (Table 19.1 to Table 19.3)

Table 19.1 : Germination percentage under different methods of planting

Treatments	*On-station*			*On-farm*		
	2004	*2005*	*Mean*	*2004*	*2005*	*Average of 10 locations*
Farmers method	76.18	95.75	86.60	86.40	85.60	86.00
Refined method	94.76	96.62	95.90	96.50	93.10	94.80
Scientific method	98.73	98.33	97.57	90.60	89.90	90.25
$CD_{0.05}$	9.35	NS	2.05	3.72	NS	2.32

(*b*) With respect to yield i.e. number of tuber per plot and number of tubers per ha, no significant difference in all the three methods of planting was found. However, the weight of tuber per plot and per hectare was found significantly (at 5% level) higher in the traditional method thereby reflecting that there were larger sized tubers in this method (Table 19.4).

(*c*) However, the ratio of number and weight of the seed planted and produce obtained was found significantly higher (Table 19.4) in the scientific (7.38 and 7.70, respectively) as well as the refined method (7.87 and 9.60, respectively) of potato planting as compared to the farmers' traditional method (5.96 and 6.80, respectively).

(*d*) The on-farm experiments also showed the similar trends as indicated in parentheses.

(*ii*) Farmers' perception on the refined method of potato planting

(*a*) Less seed rate in the refined method. Therefore costly seed materials could be saved and utilized for alternative purpose.

Table 19.2 : Comparative performance of yield related attributes (plant height) under three different methods of planting

Treatments	*Plant height*								
	45 DAP			*60 DAP*			*90 DAP*		
	2004	*2005*	*Mean*	*2004*	*2005*	*Mean*	*2004*	*2005*	*Mean*
Farmer's method	48.57(45.57)	32.98(30.98)	38.58(36.50)	60.90(57.50)	37.00(35.00)	44.84(42.50)	86.40(83.40)	45.58(43.50)	63.67(62.50)
Refine method	42.80(40.80)	38.03(34.03)	40.53(39.25)	57.97(55.00)	42.51(40.25)	47.48(45.70)	80.40(78.00)	49.64(48.60)	64.04(63.40)
Scientific method	45.40(42.20)	37.31(35.31)	41.25(38.20)	59.33(56.20)	46.45(44.40)	50.57(48.60)	75.43(76.50)	55.00(50.00)	63.85(62.50)
$CD_{0.05}$	3.50(2.90)	NSNS	NSNS	NSNS	8.59(5.62)	4.13(3.12)	4.95(3.73)	6.69(4.07)	NSNS

Figures in parentheses indicate results on farmers' field

Table 19.3 : Comparative performance of yield related attributes (Number of branches) under three different methods of planting

Treatments	*Number of Branches*								
	45 DAP			*60 DAP*			*90 DAP*		
	2004	*2005*	*Mean*	*2004*	*2005*	*Mean*	*2004*	*2005*	*Mean*
Farmer's method	3.91(3.70)	3.63(3.50)	3.76(3.60)	3.90(3.80)	4.51(4.30)	4.30(4.00)	4.63(4.50)	4.77(4.60)	4.76(4.50)
Refine method	3.86(3.80)	5.79(5.50)	4.97(4.90)	4.20(4.00)	6.00(5.80)	4.96(4.80)	4.94(4.70)	6.11(6.00)	5.76(5.50)
Scientific method	3.66(3.50)	6.29(6.00)	5.66(5.67)	4.30(4.10)	6.51(6.30)	5.53(5.20)	4.83(4.70)	6.49(6.20)	5.59(5.40)
$CD_{0.05}$	NSNS	2.09(1.90)	0.60(0.90)	NSNS	1.58(1.80)	0.73(1.20)	NSNS	1.43(1.30)	0.52(0.07)

Figures in parentheses indicate results on farmers' field

Table 19.4 : Multiplication rate (both number and weight) under different treatments

Treatments	*Number of tubers*			*Weight of tubers*		
	2004	*2005*	*Mean*	*2004*	*2005*	*Mean*
Farmer's Method	5.70(5.40)	6.20(6.00)	5.96(5.70)	7.30(7.20)	7.20(7.00)	6.80(6.70)
Refined Method	8.50(8.30)	7.20(7.10)	7.87(7.60)	9.50(9.10)	11.25(10.10)	9.60(9.40)
Scientific Method	8.10(8.00)	6.70(6.60)	7.38(7.40)	9.00(8.90)	8.70(7.60)	7.70(7.50)
$CD_{0.05}$	2.08(1.94)	NSNS	0.37(0.48)	0.52(0.47)	2.92(3.42)	0.07(0.50)

Figures in parentheses indicate results on farmers' field.

(*b*) Ease in intercultural operations in the refined method because of interplant and inter row spacing thereby facilitating the interculture operation, top dressing, etc.

(*c*) Refined method was found compatible with existing land structure as no extra alteration in the land structure was required. They also felt that the recommended scientific method of potato planting could not be adopted as such because of the existing ***'nur-bun'*** method of potato planting which is best considering the slopy land and heavy rainfall in the area.

(*d*) Convincing economics because of better return and B: C ratio in the refined method (Table 19.6).

Table 19.5 : Yield and number of tubers obtained from different treatments

Treatments	*Number of Tubers/Plant*			*Number of Tubers/ha*			*Weight of tubers/ plant (g)*			*Weight of tubers/ ha (tones)*		
	2004	2005	Mean	2004	2005	Mean	2004	2005	Mean	2004	2005	Mean
Farmer's method	5.70 (6.00)	6.20 (5.06)	5.96 (5.06)	705824 (741115.2)	775000 (790500)	745000.00 (772565.00)	507.50 (469.43)	449.85 (427.35)	478.50 (448.59)	36.54 (34.72)	36.50 (34.68)	35.70 (33.90)
Refine method	8.50 (8.25)	7.20 (8.30)	7.87 (8.20)	704448 (739670.4)	599997 (614996.9)	655830.71 (680096.49)	660.00 (610.50)	679.14 (645.18)	669.50 (627.66)	31.68 (30.10)	37.50 (35.63)	34.66 (32.93)
Scientific method	8.10 (7.75)	6.70 (5.91)	7.38 (6.83)	676074 (709877.7)	558331 (572289.3)	614997.54 (572289.3)	339.36 (313.90)	305.29 (290.02)	330.07 (309.44)	29.50 (28.02)	29.00 (27.55)	28.93 (27.48)
$CD_{0.05}$	2.08 (2.00)	NS NS	0.37 (1.20)	161376 (169444.8)	NS NS	107916.42 (111909.33)	110.90 (102.58)	109.43 (103.99)	29.27 (27.44)	0.38	7.40	2.07

Figures in parentheses indicate results on farmers' field

Table 19. 6 : Economics of different planting methods

Treatments	*Cost of production (Rs/ha)*			*Marketable yield (t/ha)*			*Gross income (Rs/ha)*			*Net profit (Rs/ha)*			*B : C Ratio*		
	2004	2005	Mean	2004	2005	Mean	2004	2005	Mean	2004	2005	Mean	2004	2005	Mean
Farmer's method	74445	74445	74445	31.06 (29.50)	31.23 (29.67)	31.05 (29.50)	186360 (177000)	187380 (178020)	186900 (177000)	111915 (102555)	112935 (103575)	112455 (102555)	1:1.50 (1:1.4)	1:1.52 (1:1.4)	1:1.51 (1:1.4)
Refine method	64441	64441	64441	26.93 (25.58)	32.66 (31.03)	30.14 (28.63)	161580 (153480)	195960 (186180)	178800 (171780)	97139 (89039)	131519 (121739)	114359 (107339)	1:1.51 (1:1.4)	1:2.10 (1:1.9)	1:1.77 (1:1.7)
Scientific method	50445	50445	50445	25.07 (23.82)	25.37 (24.10)	25.16 (23.90)	150420 (142920)	152220 (144600)	151320 (143400)	99975 (92475)	101775 (94155)	100875 (92955)	1:1.98 (1:1.8)	1:2.01 (1:1.9)	1:2.0 (1:1.8)
$CD_{0.05}$	–	–	–	**0.95 (0.90)**	**6.50 (6.18)**	**1.80 (1.70)**	–	–	–	–	–	–	–	–	–

Figures in parentheses indicate results on farmers' field

CHAPTER 21

Assessment and Refinement of Traditional Method of Late Blight Management in Meghalaya

The on-form trial was conducted to assess the effect of fungicidal spray on duration and yield of most popular potato cultivar in low land (i.e. Kufri Jyoti) and to assess the relative economics of various treatments.

The antecedents supporting trials were as follows :

1. Large area under low land potato cultivation (Approx. 4-5 thousand. ha) in Meghalaya, 2. Potato-Paddy: prominent cropping sequence in such land, 3. Potato is grown in last Feb.- last May to June (approx. 95 days crop) in such land and no plant protection is applied and subsequently, paddy is grown in last June to Nov - Dec., 4. Kufri Jyoti is preferred cultivar in such cropping sequence because of late blight susceptibility but early maturity, 5. Therefore, scope for saving the crop for another 10-15 days by fungicide spray without disturbing the cropping sequence.

Methodology adopted

1. Trial was conducted during summer season (2004 and 2005), 2. Planted on 24th Feb. & harvested on 20th June 2005, 3. Two methods of planting *viz.*, Farmers' *(Bun)* method and Scientific method were taken, 4. Two Treatments namely, sprayed (one prophylactic spray of Indofil + one spray of Ridomil) and control (no spray) were taken, 5. Experiments were laid down in RBD with four replications of each treatment and 6.Potato Cv. used was Kufri Jyoti.

Findings

(*a*) In case of delayed incidence of late blight i.e. when temperature is above or equal 25° C and RH is less than 80%, one spray of Indofil + Ridomil will be beneficial for:

1. Yield advantage : up to 48% (Table 20.1)
2. Multiplication rate : 4.4 to 6.6 (Table 20.2)
3. C : B ratio : 1:1.6 to 1:2.6 (Table 20.2)
4. No disturbance in cropping sequence and
5. Particularly applica*ble for low land production situation with Cv : Kufri Jyoti.*

(*b*) In case of timely incidence of late blight i.e. when temperature is 10-20° C and RH is above 80%, farmers' traditional method of early potato planting in low land is valid method for late blight management. However, the above experiment requires further confimation on time and space to mark the findings more valid and replicable.

Table 20.1 : Performance of various treatments for late blight management in low land potato cultivation under different cultivation methods (Year 2004 and 2005)

S.N.	*Treatment*	*Seed Rate (q/ha)*	*Crop Duration (Days)*	*Yield (q/ha)*	*Marketable Yield (q/ha)*	*Yield Advantage (5)*	*No. of Days Increase*
1	Farmersmethod (control)	50.00	95	219.09(143.20)	186.90(121.72)	-	-
2	Farmers method (sprayed)	50.00	110	324.70(180.80)	275.50(144.64)	47.00(26.25)	15
3	Scientific method (control)	33.33	95	216..90(187.20)	180.60(149.76)	-	-
4	Scientific method (sprayed)	33.33	110	281.00(210.40)	281.00(168.32)	47.94(12.39)	15
	CD at 5 %			71.6 (20.6)	61.0 (31.2)		

Figures in parentheses indicates the results obtained in 2005

Table 20.2 : Economics of fungicidal spray for late blight management in low land potato cultivation under different cultivation methods (Year 2004 and 2005)

S.N.	Treatments	Seed Rate (q/ha)	Multi-rate	Cost of Production (Rs/ha)	Gross Income (Rs/ha)	C : B	Cost Increase (%)	Income Increase (%)
1	Farmers method (control)	50.00	4.4(2.9)	70000.00	112140.00(79118.00)	1:1.6(1:1.3)	-	-
2	Farmers method (sprayed)	50.00	6.5(3.6)	74445.00	165300.00(94016.00)	1:2.2(1:1.3)	6.35	47.40(18.83)
3	Scientific method (control)	33.33	5.9(5.6)	60000.00	108360.00(97344.00)	1:1.8(1:1.6)	-	-
4	Scientific method (sprayed)	33.33	6.6(6.3)	64445.00	168600.00(100992.00)	1:2.6(1:1.6)	7.41	55.59(3.75)
	CD at 5%		NS					

Figures in parentheses indicates the results obtained in 2005

CHAPTER 22

On-farm Participatory Assessment of TPS Hybrids in Meghalaya

The evaluative trials of TPS hybrids, TPS/C-3 and 92-PT-27 were conducted on the research station as well as farmers' fields during the autumn and summer season. At on-station six grams TPS of above hybrids was utilized for ten transplanted blocks (each 20.8 m^2) measuring a total of 208 sq. mt., and same quantity of these hybrid TPS was used on ten farmers' fields in three potato growing circles of East Khasi Hills district of Meghalaya. On-farm transplanted area was 200 sq. mt. for each hybrid at each place.

Comparison was made with respect to various biological parameters, yield attributes, post-harvest behaviour and also the performance of obtained tuberlets in the subsequent season. In order to ascertain the farmers' acceptance/rejection of TPS technology demonstrated on their farms, their perception of the TPS technology was studied as the ultimate decision to adopt a technology as well as the rate of adoption of the technology is influenced by farmers' perception of the technology. In addition, the constraints perceived by them in large scale adoption of TPS were also ascertained.

Findings of the study (Table 21.1) revealed better extent of seedling germination (80-90%), lesser transplants mortality (1-2%) and lesser time taken (30-35 days) in making seedling ready for transplanting on research station as compared to the farmers' field where these figures were 60-70 percent, 2-5 percent and 35-40 days, respectively. It was found that hybrid 92-PT-27 showed relatively better performance both at the research farm and farmers' fields. The extent of occurrence and spread of late blight in TPS was found lesser at farmers' field in comparison to other prevailing varieties of potato as well as that on the research station farm.

Table 21.1 : Comparison of performance of selected TPS hybrids on research station and farmers' field

Parameters	*Performance of the hybrids*			
	TPS C-3		*92-PT-27*	
	On station	*On farm*	*On-station*	*On-farm*
Seedling germination %	80-90	60-70	90-95	60-70
Transplants mortality (%)	1-2	3-5	1-2	2-3
Incidence of late blight in seedling crop**	4	3	7	8
Average yield (q/ha)	53 (47-60)	35 (30-40)	59 (50-68)	45(42-48)
Summer crop	134	120	140	130
Autumn crop	(120-147)	100-140	125-150	110-140
Field rottage (%)	3-5	4-5	2-4	2-3
Store rottage (%)	Nil	Nil	Nil	Nil
Subsequent crop yield fromautumn tuberlets (q/ha)*	143 (KJ = 125)*	193	160 (KJ = 146)*	200 (KJ = 125)* 200(KJ = 146)*

Figures in parentheses indicate range of yield, * Kufri Jyoti (KJ) was taken as control

** 1-9 scale for late blight where 1 is susceptible and 9 is highly resistant

However, incidence of late blight was noted more in case of TPS C3 than 92-PT-27. Average total yields obtained for summer crop were 53 q/ha (range 47-60 q/ha) for TPS C3 and 59 q/ha (range: 50-68 q/ha) for 92-PT-27 at the station farm. However, at farmer's field these figures were 35 q/ha (range: 30-40 q/ha) and 45 q/ha (range: 42-48 q/ha) for TPS C3 and 92-PT-27, respectively, where as in autumn season, the yield figures were better as evident from the table. Field rottage in both the cases ranged from 2 to 5 percent with almost nil rottage in the store. Findings showed that there was considerable yield gap at the farmer's field.

Further, from the obtained tuberlets, 50 kg was planted in the subsequent summer season at the station farm and farmers' field to assess the performance of tuberlets. Cultivar Kufri Jyoti was treated as control in both the cases. Again, the incidence of late blight was noted lesser in tuberlet crop as compared to Kufri Jyoti both at farmers' field and on station. However, tuberlet crop showed lesser late blight incidence at farmer's field as compared to the station farm. Average yields obtained were 143 q/ha (Kufri Jyoti = 125 q/ha) and 193 q/ha (Kufri Jyoti = 146 q/ha) on the research station and farmers' fields, respectively for TPS C3 tuberlet,whereas for 92-PT-27, the corresponding figures were 160 q/ha (Kufri Jyoti: 125 q/ ha) at the station farm and 200 q/ha (Kufri Jyoti: 146 q/ha) at the farmer's field. Therefore, the comparative yield advantages from TPS tuberlets were computed to the extent of 12.58 percent and 24.35 percent against Kufri Jyoti on research station and farmers' fields, respectively for TPS C 3. On the other hand, still more yield advantage was recorded for 92-PT-27 (against Kufri Jyoti) tuberlets both at the station farm (21.87%) and farmer's field (27%). Findings also revealed that farmers preferred 92-PT-27 to TPS C3 because of the oval shape of its tuber and yellow flesh colour.

Based on the above findings and the experiences, it is recommended for conducting large number of such on-farm demonstrations and regular training on TPS technology particularly on nursery management, TPS agronomy, etc for the potato growers in rural Meghalaya in order to bridge the yield gap at research station and farmer's field. It was evident that TPS technology offered better yield advantage over the popular potato cultivar in the state. It is also suggested to take TPS seedling crop at smaller scale in autumn season and to utilize the obtained tuberlets as quality planting material in the subsequent cropping season to enhance the production level and to minimize the cost of cultivation.

Findings as presented in Table 21.2 revealed that majority of the potato farmers (65%) perceived TPS technology as more labour demanding in comparison to the other crops. Similarly, extent of seedlings and transplants mortality was perceived to be similar to that of other vegetable crops by majority of the farmers (above 40%). Incidence and spread of late blight in TPS crop was perceived comparatively to a lesser extent than the potato variety prevailing in the study areas by 60 percent of the potato growers. However, with regard to insect-pest infestation in TPS crop, about 62.50 percent of the farmers perceived it usual as the crop of other potato varieties. About 45 percent of the farmers agreed that potato crop through TPS required more care.

Table 21.2 : Farmers' Perception of TPS Technology

S.N.	*Parameters*	*Response*	*Rank*	*Percentage*
1.	Labour requirements :	Very high	III	12.50
	(a) Management of nursery	More than normal	I	65.00
		Usual	II	22.50
	(b) During transplanting	Very high	II	15.00
		More than normal	I	57.50
		Usual	III	27.50
2.	Mortality :	More	III	27.50
	(a) Seedling mortality	Usual	I	42.50
		Less	II	30.00
	(b) Transplants/crop mortality	More	II	42.50
		Usual	I	57.50
		Less	III	10.00
3.	Disease incidence and spread (esp. Late blight)	More	III	7.50
		Usual	II	32.50
		Less	I	60.0
4.	Insect pest infestation	More	III	10.00
		Usual	I	62.50
		Less	II	27..50
5.	Care of the crop required	More	I	45.00
		Usual	II	37.50
		Less	III	17.50
6.	Cost of cultivation	Cheaper	I	60.00
		Moderate	II	22.50
		Costly	III	17.50
8.	Overall rating of TPS crop	Superior	I	52.20
		No Difference	II	30.00
		Inferior	III	17.50

Beneficiary farmers also expressed that the most commonly perceived constraints in the large-scale adoption of TPS technology were delicate and succulent seedlings (90.00%), intensive care required at seedling stage (82.50%), more labour demand at the time of transplanting (80.00%) and also the irregular tuber shape (82.50%). The less frequently perceived constraints were incidence of late blight at nursery stage (65.00%), intensive care required for 2-3 days of transplanting (65.00%) and more variation in the shape of tuberlets in second generation (70.00%) (Table 21.3)

Table 21.3 : Farmers' perception of the constraints in large scale adoption of TPS technology

S.N.	*Crop Stage*	*Constraints*	*Rank*	*Percentage*
1.	Nursery stage	1. Intensive care (weeding, watering, etc.) is required	II	82.50
		2. Delicate and succulent seedlings	I	90.00
		3. Incidence of late blight	III	65.00
		4. Damage by ants	IV	32.50
		5. Damage by rains in summer	IV	35.00
2.	Transplanting stage	1. More labour requirement	I	80.00
		2. Mortality of transplanted seedlings	III	50.00
		3. Intensive care for 2-3 days of transplanting	II	65.00
3.	Transplanting to harvesting	1. Incidence of late blight	I	32.50
		2. Infestation of insect pest	II	17.50
4.	Post harvest stage	1. Irregular shape of tuberlets	I	82.50
		2. More variation in second generation	II III	70.00

Table 21.4 : Economics of tuberlet/seedling tuber production using TPS (92-PT-27; Year (2000-2001) in Meghalaya

Particulars	*Summer season*	*Autumn season*	*Average*
1. Yield obtained (q/ha.)	59	140	99,50
2. Total cost of production (Rs/ha.)	16,000	16,000	16,000
3. Total return obtained* (Rs)	28,320	67,200	47,760
4. Net profit (Rs/ha.)	12.320	51,200	31,760
5. Return to investment(Rs)	1.77	4.20	2.36
6. Profit : cost	0.77	3.20	1.97

*Considering 80% of the yield as the seed grade tuberlets and assumed price of quality seed at the subsidized rate of Rs. 600/q

Further, the relative economics of tuberlet production (Table 21.4) was also found in favour of autumn season wherein the return to investment (Rs) and profit: cost were 4.20 and 3.20 respectively, as compared to the respective figures of 1.77 and 0.77 in the summer season. Moreover, the tuberlet produced in the autumn season could be utilized as the quality planting material with right physiological specifications in the subsequent summer season, whereas, summer produced tuberlets can not be utilized in the immediate autumn season and hence, farmers may have to wait for complete one year for its utilization. Singh and Jee (1990) have also shown higher net return for seedling tubers (Rs. 19,552/ha) and seedling transplants (Rs. 19,174/ha) for a variety under Patna condition.

Based on the above findings, it can be concluded that despite perceiving TPS technology as labour intensive, requiring more care, most of the farmers perceived it to be cheaper for potato production and accorded overall rating of TPS technology as superior to conventional method of potato production. This clearly indicate the possibility of better acceptance of TPS technology in Meghalaya, therefore, it is recommended that more numbers of such on-farm farmer's participatory demonstrations for the potato growers should be conducted in the state. Also, extensive training programme on TPS technology for the state extension functionaries ought to be regularly organized.

CHAPTER 23

On-farm Assessment of Bio-fertilizers Application for Potato Production

Modern input intensive agricultural practices have resulted in degradation of soil fertility and associated micro-environment. Such daunting challenges ought to be addressed on priority. Besides several research and policy factors, promotion of organic agriculture could be instrumental at farm level/ micro level. Bio-fertilizers is an important component of organic farming and is known to play a major role in supplementing the crop nutrients through biological nitrogen fixation and in stabilization at bound forms of phosphorous in soil (Bharadwaj and Gaur, 1970). These beneficial microbial inoculants are also known to help in mobilizing available micronutrients in soil and are known to secrete growth promoting substances needed by the plant for their growth (Sen and Paul, 1957). Therefore, the present investigation was designed and conducted across ten numbers of farmers field in two villages namely Marbisu and Sohiong in the east khasi hill district of Meghalaya to assess the effect of bio-fertilizers in different combinations with inorganic fertilizers.

The study consisted of three treatments namely, application of 100% inorganic fertilizer + 10t/ha FYM (T1), application of 75% inorganic fertilizers + 25% through bio-fertilizers (Azotobacter + Phosphobacterium) + 10 t/ha FYM (T2) and application of 50% inorganic fertilizers + 50% Bio-fertilizers + 10 t/ha FYM (T3). The required amount of seed, fertilizers and bio-fertilizers were supplied to all the ten different farmers who acted as the replications in two different villages. The FYM however was supplied by the farmers themselves. Planting of the experiments was done in the hilly slope according to their traditional *(nur-bun)* methods. The variety Kufri Giriraj was supplied to all the farmers in addition to other production inputs. All the agronomical and plant protection measures were followed as per the need of the crop. This experiment was conducted in the summer cropping season during the years 2002 and 2003. Farmer's perception of the findings of this on-farm trial was also ascertained.

Table 22.1 : Effect of bio-fertilizer on potato production on farmers' field in Meghalaya

Treatments	*T1 (100% NPK Through inorganic)*	*T2 (75% NPK though inorganic 25% bio-fertilizer)*	*T3 (50% NPK though inorganic + 50% bio-fertilizer)*
F1	22.50	22.50	18.50
F2	24.06	23.00	18.00
F3	24.10	25.00	20.10
F4	28.37	25.50	21.50
F5	26.00	23.50	20.40
F6	23.10	25.30	21.40
F7	24.00	24.25	20.50
F8	27.00	22.05	19.50
F9	26.37	26.00	22.00
F10	23.50	24.40	23.20
Av.	25.06	24.15	20.51
CD (0.05)	3.7		
Late blight score (1-9 scale)*	5	3	3

*1 = No incidence of late blight and 9 = 100% incidence of late blight

Findings contained in Table 22.1 revealed that the mean yield of potato tuber was significantly highest (25.06 t/ha) in the treatment application of 100% NPK through inorganic source + 10 t/ha FYM over replacement of 50% inorganic source of fertilizer through bio-fertilizers (Azotobacter + Phosphobacterium) + 50% inorganic source of fertilizer + 15 t/ FYM/ha (20.51 t/ha). However, replacement of 25% inorganic source of fertilizer through bio-fertilizers (Azotobacter +

phosphobacterium) + 75% inorganic sources of fertilizer + 15 t FYM /ha recorded at par result (24.15%ha) with the application of 100% NPK through inorganic sources of fertilizers. This may be due to favourable conditions created by bio-fertilizer with sufficient nutrient supply at early growth and development stages of the crop. It is in agreement with the findings at Sharma *et al.* (1996) and Singh (1999). Similarly, the significant decrease in yield due to application of only 50% of inorganic NPK might not fulfill the demand for nutrient to the exhaustive crop potato due to immobilization of nutrients during early growth period of the crop as well as at maximum tuberization stage.

The cost benefit ratio calculated with regards to different levels of fertilizers and bio-fertilizers recorded the same (1: 4.2) in case of 100% NPK and 75% NPK + 25% biofertilizers (Table 22.2). However, lower value was recorded (1:3.7) in case of 50% NPK + 50% bio-fertilizers. Similar results were also reported by Praharaj *et al.*, 2002.

Table 22.2 : Comparative economics of application of different level of bio-fertilizers on farmers' field

S.N.	*Treatment*	*Cost of Production (Rs/ha)*	*Gross Return (Rs/ha)*	*C : B Ratio*
1.	100% NPK through inorganic source (T1)	35,853	1,50360	1:4.2
2.	75% inorganic source + 25% organic source (T2)	34,430	1,44900	1:4.2
3.	50% inorganic + 50% organic source (T3)	33,007	1,23060	1: 3.7

Farmers' perception of this on-farm trial indicated (Table 22.3) that the technological option of replacing the chemical fertilizers with bio-fertilizers to certain extent helped in cost saving. Besides the mode of application was found to be less complex to adopt it fully. Many of the secondary and tertiary beneficiaries of this on-farm trial opined that bio-fertilizer treated potato field appeared healthy and relatively lesser infected by late blight and other diseases in comparison to the non-treated field. In addition, all the beneficiary farmers felt that the soil condition was better in bio-fertilizer treated plot for the subsequent vegetable crops in the same plot as compared to the untreated field.

Table 22.3 : Farmers perception of use of bio-fertilizer in potato crop

Sl.No	*Farmers perception*	*Rank*
1.	Cost saving	I
2.	Easy to use	II
3.	Improved soil health	III
4.	Lesser incidence of late blight	IV

Findings and experiences of this on-farm experiment helped to conclude that reducing the level of inorganic fertilizer up to 25% through use of bio-fertilizers (Azatobacter + Phosphobacterium) is more effective in relation to yield, economiçs and especially with regards to good soil health and fertility in high rainfall areas of Meghalaya.

CHAPTER 24

Technological Empowerment of Potato Growing Tribal Farmwomen

Women in Meghalaya

Women folk in the tribal state constitute about fifty percent (49.35%) of the total population. The sex ratio in the state is 975 female per thousand males. Half of the population in this state is literate (50.75), while the female literacy rate in the state is 48.42 percent.

Female population in the state constitute about 42 percent of the total workforce against the national figure of 31.56 percent. On further break up, their percentage as main workers is only 36.48 percent against 59.76 percent as marginal workers in the state. As far as participation of female work force in agriculture sector in the state is concerned, about 72 percent of the total female workers in the state are engaged in agriculture as cultivators and agricultural labourer, against 61.13 percent of total male workers. About 45.84 percent of the total cultivators in Meghalaya state are women, while the corresponding national figure is only 32.36 percent. This difference could be attributed to the social status of women in the tribal community of Meghalaya.

Thus, tribal women play a significant role in agriculture of Meghalaya state and the women belonging to the three ancient hill tribes in this state are the **'real farmers'**, the active work force engaged in agriculture rather than simply **'farmer's wives'** or **'housewives'**.

Women in Meghalaya state possess a higher socio-economic status owing to the existing matriarchal and matrilineal structure existing among the native tribal communities in the state. Thus, owing to the existing social system and the social structure, the agricultural land along with other property is transferred from mother to daughter, giving the women a right over the agricultural land and thus giving women a better say in agricultural related activities and decisions in the state. Along with all other crops, tribal women have an important role in management of potato production as well as post harvest handling in the state.

Rationale for women empowerment in potato production

Empowerment of farm women with the appropriate agricultural production technologies is being recognised by the researchers and development managers as a tool for further gearing the agricultural production scenario of India in positive direction. However, for exploiting the true potential of the improved scientific technologies, the actual work force need to be made technologically competent or in other words made technologically empowered. Technological empowerment of agricultural women work force intends their capacity building regarding improved agriculture production technologies by way of creating awareness among them and enhancing their level of knowledge and symbolic adoption about the technologies, inculcating favourable attitude towards the same and improving their operational skills regarding the correct application of these technologies in the field.

The tribal women are the predominant work force in agriculture in the state, including potato production and post harvest handling. Their role is further heightened owing to matrilineal as well as matriarchal social system. The low yield (9.5t/ha) in Potato gives adequate reasons for directing efforts towards building the capacity of actual potato producers, on various aspects of improved potato production and post-harvest practices. Hence in order to mitigate the problem of low productivity of potato crop and subsequently to ensure higher economic returns to the farm households in the state, efforts are required to be made for capacity building of tribal women farmers. In the similar view Desai and Mohiuddin (1992) in a study of three Indian states including Meghalaya also observed that building technical competencies and energising extension system to the requirements of farm women were some of the issues needing immediate attention.

In this line timely and need based training supported with on-farm demonstration was identified as the crucial qualitative input for improving knowledge, skills, and attitude of human resources thereby empowering them technologically. This finds equal applicability in case of technologically empowering of tribal farm women of rural Meghalaya with improved potato production technologies. The training intervention could be further effective if it is planned and implemented as per the crop stages.

Further, training programme directed towards technological empowerment of women farmers would fail to show its impact if it is not evaluated. It is therefore, important to assess the change in level of knowledge, attitude, symbolic adoption and retention of knowledge related to improved production technologies among the trained tribal women due to training. The present chapter, therefore, was developed in light of the above background with the following specific objectives :

1. To quantify tribal farmwomen's contribution in production and post-harvest handling of potato.
2. To build farm women's capacity on improved potato production technologies and post-harvest practices through need-based training intervention.
3. To assess the impact of training intervention on behavioural dimensions (knowledge, attitude and symbolic adoption) of tribal farm women

Review of Related Literature

(i) *National status*

Human resource is the most precious resource for any nation. It is, however, not the numerical but the qualitative strength of the people which forges a country ahead towards progress and prosperity. It is basically the development of human resources that brings about socio economic and politico cultural transformation of any society. The basic problem of any under-developed/developing country is not the scarcity of natural resources, but the under-development and limited exploitation of human resource. India is also not the exception to it. Hence, building up the human resource is one of the major tasks of the country. The United Nations Development Programme's (UNDP) Human Development Report (1990) confers with the opinion that any further significant economic development can take place only if the human resource is trained, mobilized and applied (Venkatesan, 1993). Training has been identified as the crucial input for development in all walks of life, be it agriculture, animal husbandry, industry or any other field through the development of human resource.

Though significant progress has been achieved by Indian agriculture in terms of agricultural production as well as productivity in almost all the crops, there exist wide gap between the potential yields and the yields obtained by the farmers at the field level.

The need for effective training and education was underlined as one of the steps of improved technology development and transfer with overall system approach by Johnson (1987). However, such training programmes are of limited value if proper evaluation is not carried out and impact in terms of change in knowledge, attitude and skills among trainees is not studied and benefit is drawn fro the gained experience for bringing about improvement in the future training efforts.

(a) *Gain in Knowledge due to Training*

Gain in knowledge refers to the additional information acquired by the trainees as a result of attending the training programme. In most of the studies, researchers have found increase in the knowledge level of trainees after the training (Gour and Shrivastava, 1983; Babu and Singh, 1986; Vashistha, 1987; Chauhan *et al.*, 1990). However, the extent of gain in knowledge varied from one study to another. Babu and Singh (1986) reported the average gain in knowledge about 20.28 percent, Ananthrama and Ramnathan (1990) reported it to be 62.77 percent. Similarly, knowledge gained as a result of attending the training programme was found to be 89.26 percent as reported by Fulzele (1986). In further study, Sah (1996) reported as high as 110 percent gain in knowledge of farm women owing to the training exposure. These literature reveal the differential increase in level of knowledge of trainees due to the training programme. This leads to infer that there was considerable variation in the course content, training methodology and approaches, followed by the training organisations as well as the variation in the socio economic and psycho cultural parameters of the trainees.

(b) *Retention in Knowledge due to Training*

Retention of knowledge refers to the ability to keep/retain and retrieve the knowledge learnt during the training programme. Extent of impact of any training programme also depends on the capacity of the trainees to retain the learnt information. In this line, Fulzele (1986) found a significant retention of knowledge of scientific practices among trained dairy farmers after six months of training, while Perumal *et al*. (1993) more objectively, reported 60 to 75 percent retention of knowledge among trained client. Similarly, Sah (1996) found the retention of knowledge to the extent of 87 percent among the trained dairy farm women. The variation in the retention of knowledge among trainees could be explained in terms of nature of technologies/ practices learnt, style of teaching, learning situations, etc.

(c) *Attitudinal Change due to Training*

Attitude is considered as an organised pre disposition to think, feel, perceive and behave towards a referent as cognitive object. Training intends to bring about and sustain the positive change in the attitude of the clients. Researches have indicated the positive and significant change in the attitude of trainees as a result of training programme (Babu and Singh, 1986; Swaminathan, 1989; Satyanarayan *et al*., 1994). Similarly, Kamalsen (1971) reported that 62 percent of the farmers had increased their attitude towards high yielding varieties of paddy. Development of favourable attitude of participants towards agricultural innovation as result of training was reported by Umamahesha and Channegowda (1988). More objectively, Sah (1996) found significant change of 11 percent in the attitude of dairy farm women of Haryana due to the training.

(d) Change in Symbolic Adoption due to Training

Symbolic adoption refers to the mental decision of the respondent to make full use of recommended practices. This cognitive decision stage is a vital step prior to practice adoption. Unless this mental willingness is found out and the negligence towards innovation are rectified, proper adoption may not turn out to be a reality. Training intends to rectify such negligence and thereby increase the likelihood of the adoption of the practices in coming future. Regarding its empirical evidence, several studies since sixties to nineties reported the significant impact of training on the symbolic adoption level (Singh, 1968; Vashistha, 1987; Reddy, 1989; Fulzele, 1995). More precisely, Singh and Singh (1990) reported that the increase in mean symbolic adoption score was from 33.54 to 58.78 as a result of training. Similarly, Sah (1996) found an increase of about 18 percent in the symbolic adoption level of farm women due to training. It could be hence, concluded that appropriate training strategy would positively manipulate the cognitive domain of the trainees and hence, an individual can be turned into the human resource in real sense.

(e) Diffusion Effect by Trained Human Resource

Diffusion effect as defined by the Rogers and Shoemaker (1971) is the cumulatively increasing degree of influence upon an individual to adopt or reject an innovation resulting from the increasing rate of knowledge and adoption or rejection of the innovation in the social system. Trained human resource may also facilitate to change the overt as well as covert behaviour of the other members of their social system. Hence, training could serve as the fuel to start chain reaction for the generation of human resources. Study of Vashistha (1987) reported that training had some diffusion effect on the knowledge and adoption score of the non participating farmers of the same village as that of the trained counterpart. Similar findings were also reported by Sah (1996). Though the literature available in this regard is scanty, from the above studies, however, it could be interpreted that the effective training programme has a wider influence on society, in terms of trained as well as untrained members.

(ii) International staus

The need for effective training and education was underlined as one of the steps for improved technology development and transfers will overall systems approach by Johnson (1987).

Based on the literature reviewed the in preceding subheads, it could be concluded that on all the indicators of impact, training had significant influence. The variation in the level of impact could be explained in terms of the differing parameters of both the training organisation as well as the beneficiary trainees delineated as above. Despite such variation, training could be identified as the most important means to accomplish the ends in terms of changing the behavioural dimensions of the clients. Literature also revealed that the impact was relatively more on the cognitive domain as compared to the affective dimension of behaviour of the trainees. Further, the diffusion of the learnt information from trained clients to non-trainees was also found considerable. However, there was no information in the literature reviewed from the tribal settings f Meghalaya and with respect to potato crop. This project was, therefore, an attempt to bridge the above knowledge gap.

Methodology Adopted

Sampling plan

A multistage random sampling technique was applied to draw the sample for the study. The present study was carried out in purposively selected East and West Khasi Hills district of Meghalaya, for the reason of having maximum area under potato crop in the state consequently having the highest production of potato among all the seven districts of the state of Meghalaya.

A total of 200 farm women were randomly selected from 14 selected villages from these two selected districts. Half of the sample i.e., 100 women potato growers were further selected as experimental group and the other half was selected to act as control in order to see the diffusion effect, if any, due to the training programme.

Operationalisation and Measurement of the Research Variables

(i) Participation of farmwomen in potato production and post harvest handling practices

For studying the extent of involvement of respondents in terms of activities performed, lists of 19 activities were prepared, which included six, four, four and five activities in potato production, plant protection, harvesting, post-harvest practices and marketing related activities, respectively. Similarly, in case of decision making the various aspects of potato production and storage, respondents were enquired for the listed 13 aspects which included four, six and three activities related to potato production, utilization of inputs for potato production and marketing related activities respectively. Each respondent were asked to give response as to who perform the said activity in her house on the given four options viz., man, woman, man and woman jointly, and any other.

Access and control over resources related to potato production and post harvest handling as well benefits accruing there upon was studied utilizing a appropriate schedule developed for the purpose.

(ii) Needs related to improved potato production and post harvest handling practices

Needs for the present investigation was operationalised as the training and infrastructure needs of potato growing farm women related to improved potato production and post harvest handling practices. The finally prepared interview schedules consisted of overall 10 training and 6 infrastructure need items. Response from the sampled tribal women farmers on these needs were ascertained on three-point continuum, viz., urgently needed, needed, and not needed, and the accordingly the scores of 2, 1 and 0, respectively, were assigned to them. The need index for individual item were then calculated.

Knowledge, symbolic adoption and attitude related to improved potato production and post harvest handling practices.

Gain in knowledge, the research variable for the present study were calculated by subtracting the pre- training knowledge score form the post training knowledge score.

Retention of knowledge variable for the present study was studied by assessing the knowledge retained by farm women after 30 days of training. The variable was measured by administering the same knowledge test after 30 days of completion of training and subtracting the pre- training knowledge score form the delayed test score.

For the present study, symbolic adoption was operationalised as the mental acceptance of improved potato production and post harvest handling practices and was measured by appropriate interview schedule.

Attitude for the present study was operationally defined as the tendency of individual potato growing farm women to respond favourably or unfavourably towards improved potato production and post harvest handling practices. The attitude variable was measured by using appropriate scale.

Change in the symbolic adoption and attitude score was calculated by subtracting the pre- training score form the post training score.

(iii) Diffusion effect

Diffusion effect is the cumulative increase in the degree of influence upon an individual to adopt or reject an innovation, resulting from the increasing rate of knowledge and adoption or rejection in the social system (Rogers and Shoemaker, 1971). The research variable diffusion on effect for the present study were operationally defined as the influence of trained farm women on the untrained farm women in terms of the gain in knowledge, change in symbolic adoption and attitude score. The variable was measured by administering the same knowledge, symbolic adoption and attitude schedules test after 30 days of completion of training to the untrained farm women and subtracting the pre- training scores form the delayed test scores.

Building farm women's capacity through need-based training intervention

In order to built capacity of the tribal farm women related to improved Potato production and post harvest handling practices, the following approach was utilized.

Data collection and analysis

Data pertaining to research variables namely, participation, decision making, access and control over resources and benefits, knowledge, symbolic adoption and attitude related to Improved potato production and post harvest handling practices were collected though well-structured pre-tested interview schedule by personal interview method. A

Table 23.1 : Research Design utilized for the data collection in the study

Data Collected On	*Trained Farm women (n=100)*	*Untrained Farm (n=100)*
A. Pre- training stage		
(i) Knowledge level	Yes	Yes
(ii) Symbolic adoption	Yes	Yes
(iii) Attitude	Yes	Yes
B. Training imparted	**Yes**	**No**
C. Post training stage		
(i) Knowledge	Yes	No
(ii) Symbolic adoption	Yes	No
(iii) Attitude	Yes	No
(iv) Retention of knowledge (After 30 days)	Yes	No
(v) Diffusion effect (After 30 days)	Yes	Yes

post training survey were conducted on the research variables to ascertain the change in the scores. The research design utilized for data collection for the present study is being presented in the table 23.1.

The collected data were edited, tabulated and subjected for the descriptive statistics as frequency, percentage and mean to draw meaningful conclusions. The findings of the study have been deliberated in the subsequent subheads.

Findings and Discussion

To quantify tribal farmwomen's contribution in the production and post harvest handling

Contribution of tribal farmwomen was ascertained in terms of their participation in performing various activities, decision-making and access and control of resources as well as benefits related to potato production and post harvest handling. The further details on above aspects of farmwomen contribution in production and post-harvest handling of potato have been presented as below.

Participation of tribal women farmers in the potato production, protection, post harvest and marketing related activities/practices

(i) Participation of tribal women farmers in the potato production practices

The findings as contained in the Table 23.2 showed that participation of women farmers was higher in activities like FYM application in the field (53%), application of chemical fertilizers (51%), potato planting (49%) and selection of seed for planting (47%). A considerable percent of respondents reported that activities like FYM application (27%), potato planting (35%), seed selection (38%) and fertilizer application in fields (31%) was carried out jointly by women and men together. As high as 88 percent of the respondents reported that weeding and earthing operations were carried out exclusively by women. Men, on the other hand were reported to participate in the above mentioned activities by a lower percentage of the respondents. Land preparations for potato planting was the operation that was reported to be carried out jointly by men and women farmers by majority of the respondents (59%).

Thus it is concluded that majority of the activities related to potato production were carried out primarily by women farmers either singly or jointly with men in this hilly tribal state of Meghalaya.

Table 23.2 : Percentage distribution of potato growing women farmers' participation in potato production practices (n = 200)

S.No.	Practices	Women	Men	Jointly	Others
1.	Land preparation	10	21	59	10
2.	**FYM application**	53	13	27	07
3.	Potato planting	49	09	35	06
4.	Seed selection	47	–	38	–
5.	Seed treatment	–	–	–	–
6.	Fertilizer application	51	15	31	03
7.	Weeding and earthing up	88	05	03	04

(ii) Participation of tribal women farmers in the plant protection practices

Data presented in Table 23.3 revealed that among the three identified plant protection related activities, application of insecticide (82%) and application of fungicide (67%) were carried out by men while, rouging of diseased and off type of plants was carried out jointly (17%) by men and women potato growers.

Table 23.3 : Percentage distribution of potato growing women farmers' participation in potato protection practices (n = 200)

Sl.	Practices	Women	Men	Jointly	Others
1.	Application of fungicide	23	67	10	–
2.	Application of insecticide	–	82	–	11
3.	Rouging of diseased and off-type plants	09	–	17	–

Participation of men was thus found higher in plant protection related activities as compared to women in Meghalaya.

(iii) Participation of tribal women farmers in harvesting and post-harvest related activities

The result presented in table 23.4 revealed that majority of the respondents reported that the practices like harvesting of potato (66%), sorting and grading of potato (52%) and also the storage of potato (45%) were done by the women farmers exclusively. About 28 percent of the respondents observed that exclusively men performed storage of harvested produce and while 27 percent reported that this operation performed by men and women jointly. Carrying harvested produce from fields to store was reported to be performed by men exclusively (76%).

Table 23.4 : Percentage distribution of potato growing women farmers' participation in harvesting and post-harvests related practices (n = 200)

Sl.	*Practices*	*Women*	*Men*	*Jointly*	*Others*
1.	Harvesting of potato	66	07	27	–
2.	Sorting and grading of potato	52	10	38	–
3.	Carrying of harvested produce from field to store	06	76	18	–
4.	Storage of potato	45	28	27	–

(iv) Participation of tribal women farmers in potato marketing related activities

Potato marketing related activities majority of the respondents revealed that activities like purchase of fungicide, purchase of fertilizers and purchase of insecticide were primarily done by the men farmers i.e., 50, 48 and 42 percent, respectively. Disposal of harvested potato was carried out by men (32%), women (30%) and by men and women jointly (31%).(Table 23.5).

It is evident from the above results that no clear-cut differentiation in gender roles existed as far as potato marketing related activities is concerned in the state.

Table 23.5 : Percentage distribution of potato growing women farmers' participation in potato marketing related practices (n = 200)

Sl.	Practices	Women	Men	Jointly	Others
1.	Purchase of seed	5	15	53	–
2.	Purchase of insecticides	38	42	20	–
3.	Purchase of fertilizers	33	48	09	–
4.	Purchase of fungicides	31	50	19	10
5.	Disposal of harvested potato	30	32	31	07

The results revealed that tribal women farmers in Meghalaya are involved in performing most of the works related to production and post-harvest handling of potato. The male counterpart, on the other hand, were more involved in the activities like plant protection measures and activities like carrying of harvested produce from field to store. In other words, tribal women were found to be the major operator in almost all the activities related to production and post harvest handling of potato in rural Meghalaya.

Decision making pattern by the tribal farm women in the potato production, protection, harvesting and post harvest handling

(i) Decision making pattern in potato production practices

Findings presented in Table 23.6 revealed that women had major role in deciding the activities like land to be utilized for potato cultivation and number of nurs (raised beds) as reported by 52 and 49 percent of the respondents, respectively. Joint decisions by men and women farmers on aspects like planting time and potato varieties to be planted as reported by 68 and 55 percent respectively.

It could be concluded from the data that majority of the decisions related to potato production were either taken by women alone or in consultation with their male counter parts. Male farmers exclusively were less involved in the above said decisions.

Table 23.6 : Percentage distribution of potato growing women farmers on the basis of decision making pattern related to potato production practices (n = 200)

Sl.	*Practices*	*Women*	*Men*	*Jointly*	*Others*
1.	Land to be utilized for potato cultivation	52	08	40	–
2.	Number of nurs (raised beds) to be utilized for potato cultivation	49	10	41	–
3.	Potato varieties to be planted	24	18	55	03
4.	Planting time	18	15	68	–

(ii) Decision making pattern related to utilization of potato production inputs

The decision making pattern in case of utilization of potato production inputs were observed similar to the decision making pattern related to potato production practices. It could be seen from the Table 23.7, that decisions related to the amount and sources of other production inputs to be utilized, labour to be utilized were taken exclusively by women as reported by 60 and 46 percent of the respondents, while decisions on seed source to be utilized (53%), time of application of fungicides and insecticides (55%) were predominantly jointly decided by men and women farmers.

Table 23.7 : Percentage distribution of potato growing women farmers on the basis of decision making pattern related to utilization of potato production inputs (n = 200)

Sl.	*Practices*	*Women*	*Men*	*Jointly*	*Others*
1.	Seed replacement	05	07	10	07
2.	Seed treatment	–	–	–	09
3.	Seed source to be utilized	28	11	53	08
4.	Amount and sources of other production inputs to be utilized	60	15	25	–
5.	Labour to be utilized	46	19	35	–
6.	Time of application of fungicides and insecticides, etc.	13	32	55	–

Most of the decisions related to utilization of potato production inputs were taken either by women alone or jointly by men and women. Men exclusively had less role in these decisions.

The findings thus helped to infer that tribal women farmers were not only the active actor in the operation of potato production, protection, harvesting and post harvest handling of potato, they have the active decision makers in the operations of entire set of above activities.

Access and control over the production resources and the benefits accrued from the farm operations

Access and control over the farm resources and the benefit are the important criteria for women's socio-economic empowerment in any agricultural production system. The findings is given in the Tables 23.8 and 23.9 showed that regarding use of the production resources like land, labour, farm implements and seed both men and women farmers had the equal access. However, women were found to have edge over men counterpart in having control over the resources. This may be because of the fact that social system in the rural Meghalaya is basically matrilineal and matriarchal in nature.

Table 23.8 : Percentage distribution of tribal women farmers on the basis of access and control over production resources (n = 200)

Sl.	*Production resources*	*Access*	*Control*
1.	Land	MW (100)	W (80), M (20)
2.	Labour	MW (100)	(20)W (81), M (19)
3.	Farm implements	MW (100)	MW (100)
4.	Seed	MW (100)	MW (90)

M = Men; W = Women

Also with respect to the access and control over the farm benefits, it could be seen that both men and women had equal extent of access on farm income, seed potato and table potato. And again, the control over these farm benefits rest mainly with women farmers.

Table 23.9 : Percentage distribution of tribal women farmers on the basis of access and control over farm benefits (n = 200)

Sl.	*Production resources*	*Access*	*Control*
1.	Farm income	MW (100)	W (83), M (17)
2.	Seed potato	MW (100)	W (75), M (25)
3.	Table potato	MW (100)	W (78), M (22)

The findings thus lead to conclusion that the existing social system allows women potato grower having enough role in almost all aspect of potato production practices in rural Meghalaya. However, in order to make them technologically empowered, it was essential to asses their existing level of knowledge, symbolic adoption and attitude towards improved potato production practices. Accordingly, their need perception related to training on the same was also ought to be ascertained.

Existing level of knowledge, symbolic adoption and training needs of women potato farmers towards improved potato production practices and post harvest handling

(i) Existing level of knowledge and symbolic adoption

The data presented in Table 23.10 reveal that farm women possessed about 42 and 50 percent of overall knowledge and symbolic adoption level respectively with regard to improved potato production practices and post harvest handling. Aspect wise, it was found that they had more existing knowledge in the post harvest handling of potato (43%) followed by potato production (42%) and plantprotection (about 40%). On further analysing the data with respect to seven important aspects related to improved potato practices (Table 23.11), it was observed that the sampled tribal farm women possessed least knowledge with regard to scientific potato planting, followed by management of brown rot. While on the other hand, the mean knowledge score with regard to management of late blight disease was found to be highest i.e., 47 percent among all the seven major aspects studied, followed by improved post harvest handling of potato (43%) and potato seed management (40%).

The poor knowledge level of farm women with regard to management of brown rot and also scientific planting method could be attributed to the prevailing *'nur-bun'* system i.e., potato cultivation on raised beds in Meghalaya state.

Table 23.10 : Existing knowledge and symbolic adoption level of tribal women farmers on major aspects of scientific potato production practices

Sl.	*Aspects*	*Knowledge level*	*Symbolic adoption level*
1.	Potato Production	42.42	50.00
2.	Potato protection	39.86	46.50
3.	Post harvest handling	43.13	55.00
	OVERALL	41.83	50.50

Table 23.11 : Existing knowledge and symbolic adoption level of tribal women farmers on major areas of scientific potato production practices

Sl.	*Aspects*	*Knowledge level (%)*	*Symbolic adoption level (%)*
1.	Seed management	40.00	55.25
2.	Planting method	20.50	69.00
3.	Balanced use of chemical fertilizersManagement of late blight	35.00	52.38
4.	Management of brown rot	47.24	53.35
5.	Management of insect pest	25.13	49.67
6.	Post harvest handling of potato	36.00	40.45

A perusal of the data (Table 23.10) on symbolic adoption reveals that the sampled women tribal farmers had 50 percent of overall symbolic adoption regarding improved potato production and post harvest handling practices. Data presented in table 23.11 relating to seven major aspects reveal that symbolic adoption level was highest with regard to scientific planting method (69%). Seed management (55%), management of late blight (53%) and a balanced use of chemical fertilizer (52%) were the aspects having better symbolic adoption level.

Least symbolic adoption level i.e., 40 % was found among the respondents with regard to management of potato pest.

(ii) Training need perception of the women farmers

Training needs of tribal potato growing farm women of Meghalaya related to improved potato production and post harvest handling practices was studied in all the major areas i.e., potato agronomy, plant protection and post harvest handling. These are dealt in detail in the following sub heads:

Training needs related to potato agronomy

From Table 23.12, it could be observed that among all the five training need items relating to potato agronomy, training need was perceived highest in scientific method of planting (56.67%), followed by balanced use of chemical fertilizer (53.75%) and land preparation (41.67%). Training need on items like soil testing for fertility status (21.67%) and application of FYM (22.92%) was perceived in a lower intensity.

The high training need with respect to scientific method of planting could be explained on the basis that tribal potato farmers in the state follow the traditional potato planting in pits dug on raised beds, which is slightly labour intensive. Another characteristic feature of potato cultivation in Meghalaya is the heavy use of FYM along with minimum use of chemical fertilizer, which also explains the high training need index with respect to the item like balanced uses of chemical fertilizer along with a low need with respect to application of FYM.

Training needs related to plant protection measures

Among the four items studied pertaining to plant protection, training need was perceived highest in management and control of late blight with need index of 72.08 percent, followed by management of brown rot for which need index was 62.92 percent by the tribal farmers of Meghalaya. While for item like chemical treatment of seed before planting the seed index were only 44.17 percent. The training need was found lowest on the item of insect pest identification and control (Table 23.12).

Late blight is a regular occurring disease in the potato crop in the state that affects both the summer and autumn crop in devastating manner. This could be the reason for the item management and control of late blight acquiring highest priority by respondents for training. Brown rot is another serious problem affecting potato in the field as well as in store. Insect pest identification in the state is less apparent except during dry warm season in comparison to the devastation by diseases. This could be the probable reason for the response of the sampled farm women on the training need item of insect–pest identification and control.

Training needs related to Post Harvest handling

Table 23.12 : Training needs perception of sampled tribal women potato growing farmers regarding improved potato production and storage practices (n = 200)

Sl.	*Training Need Area*	*Training Need Index*	*Overall Rank*
A.	**Potato Agronomy**		
1.	Land Preparation	41.67	III
2.	Scientific method of Planting.	56.67	I
3.	Soil testing for fertility	21.67	IV
4.	Application of FYM	22.92	V
5.	Balanced use of chemical fertilizer	53.75	II
B.	**Plant Protection Measures**		
1.	Chemical treatment of seed before planting	44.17	III
2.	Management and control of late blight.	72.08	I
3.	Management and control of brown rot	62.92	II
4.	Insect pest identification and control	41.67	IV
C.	**Post Harvest Handling of Potato**		
1.	Grading and sorting of potato	15.42	II
2.	Proper storage of potato	46.25	I

Grading & sorting and the storage of potatoes were the only two identified training need items in the post harvest practices. On closer look at the training need index obtained for this area, it could be seen that the index were lower as compared to the need index of items in other area. Among the two post harvest handling items training need of respondents was found to be higher for item on proper storage of potato (46.25%) and for grading and sorting of potatoes the training need index was found to be low i.e. 15.42 percent only (Table 23.12).

Potatoes in Meghalaya are mainly stored in traditional methods, which cause heavy rottage and damage. However, the tribal farmers practice grading and sorting of potatoes based on their own perception of size of tubers. These facts could be the probable reasons for the items of proper storage of potato, getting a high training need index and grading and sorting item receiving lesser priority.

Infrastructure need perception of women farmers:

Availability of adequate infrastructure facilities like shops for supply of fertilizer, transportation, motor able roads, etc is one of the important factors that may determine the acceptance of recommended/improved potato production practices among the tribal farmers of the state who are operating in a hilly and difficult terrain. An attempt was therefore made to analyze the infrastructure requirement in the study locale as perceived by the potato growers.

A perusal of Table 23.13, showed that infrastructural facilities like cooperative store at approachable distance for availability of fertilizer and agency for making market information available to the farmers on time received highest priority by the potato growers in the state with 66.25 and 59.58 percent of average infrastructure average need index, respectively.

Government shop for supply of fungicide, insecticide at appropriate distance and provision for well-knit seed supply agency was the other infrastructure facilities were accorded high priority by the farmers with 58.75 and 54.17 percent need index. Potato storage related items like facility for cold storage and facility for country made storage godowns, were the items that had low need index i.e. 46,25 and 40.00 percent, respectively.

Table 23.13 : Infrastructure needs related to Potato production, storage and marketing (n = 200)

Sl.	*Infrastructure need Area*	*Need Index*	*Overall Rank*
1.	Provision for well-knit seed supply agency	54.17	IV
2.	Cooperative store for availability of fertilizer at appropriate distance	66.25	I
3.	Govt. shop for supply of fungicides, insecticides etc at appropriate distance	58.75	III
4.	Facility of cold storage	46.25	V
5.	Facility of country made storage (ware house/godowns)	40.00	VI
6.	Market Intelligence Agency	59.58	II

The high priority assigned to the establishment of cooperative storage for supply of fertilizer got support from the high training need on balanced use of chemical fertilizer like wise lower infrastructure need in storage related items also gets support from the lower training need as perceived by the respondents. One of the reason for according the storage related items like establishment of cold storage as well as country made storage structure a lower priority coupled with high need in establishing agency to market intelligence is that the potato growing tribal farmers finds a ready market of their potato produce within the state as well as markets in the neighbouring north eastern states.

Owing to the fact that potato crop in the state is grown in summer, which is off-season for the neighbouring states in plain areas, thus, the need for storage is less in the state and whatever short term storage for 3-4 months is needed is to get a higher market price that also could be minimized with proper market intelligence of the domestic market as well as the potato market in the neighbouring states.

To build farm women's capacity on improved potato production and post harvest practices through need based training interventions

In order to build farm women's capacity, 100 potato growing tribal farm women were imparted training on the identified and prioritized training need area related to potato production and post harvest handling. Training needs were identified in almost all the areas of improved potato production and post harvest practices of potato.

In order to make the training more effective, it was imparted in two phases i.e., pre planting and full grown crop stage. In addition, theoretical lectures as well as practical hands on experience were provided to the participants. Details about the training are presented in the following Table 23.14.

Table 23.14 : Details of training module developed and Implemented

S. No.	*Crop Stages*	*Month of training*	*Area covered*	*Mode*
1.	Pre planting stage	Feb-March 2003	Seed management Fertilizer management Land preparation Planting method, TPS nursery raising	Theoretical lectures and Practical sessions
	Full Grown crop stage	May-2003	Insect pest management Disease management Post harvest handling, TPS transplanting	-do-

To assess the impact of training interventions on the behavioural dimensions of women trainees

To assess the impact of training among the women trainees, comparison of pre- and post training knowledge with regard to the various behavioural dimensions like knowledge level, attitude, symbolic adoption in improved potato production and post harvest practices in potato was carried out. The details of the change in behavioural aspects are mentioned in the table 23.15.

As indicated in the Table 23.15 it was found that there was considerable change/gain in the behavioural dimensions of tribal women farmers. These changes ranged from 13.33 percent in case of change in attitude to 40.59, 42.85 and 55.38 percent in the symbolic adoption level, retention of knowledge and change in the knowledge level, respectivel.

Table 23.15 : Impact of training programme on the overall behavioural dimensions of the sampled tribal women farmers (n = 100)

S.N.	*Impact Indicators*	*Stages of Training*	*Mean Score*	*Mean Difference*	*% Change in Mean*
1.	Gain in knowledge	Before	41.83	23.17	55.38%
		After	65.00		
2.	Retention of Knowledge	Before	41.83	18.17	42.85%
		After	60.00		
3.	Change in symbolic adoption	Before	50.50	20.50	40.59%
		After	71.00		
4.	Change in attitude	Before	26.00	04.00	13.33%
		After	30.00		

It could be also seen from the Table 23.15 that the gain was more prominent in case of knowledge followed by symbolic adoption and attitude. A relatively lesser change in the attitude score of the women farmers was observed that attitude, an effective component of behaviour is a relatively permanent in nature that require repeated and strong reinforcements for bringing out significant changes.

Similarly, a lesser change in symbolic adoption score could be explained on the base that the pre-training average symbolic adoption score of farm women with regard to improved potato production and post harvest practices was found to be as high as 50.50 percent and hence the gain in was also observed lesser i.e., 40.59 percent. However, a maximum change in the knowledge score after training were recorded from a pre–training average knowledge score of 41.83 percent to 65 percent after training.

Table 23.16 : Impact of trainings programme on different behavioural aspects of tribal women potato farmers in various aspects of scientific potato cultivation

Sl. No.	Aspects	Av. Knowledge level					Av. Symbolic adoption level			Av. Attitude level		
		Pre-Training	*Post-Training*	*Gain (%)*	*Av. Knowl. level after 30 days of training*	*Retention of knowledge (%)*	*Pre-Training*	*Post Training*	*Change (%)*	*Pre-Training*	*Post-Training*	*Change (%)*
1.	Potato Production	42.42	70.00	65.00	60.00	41.44	50.00	80.00	60.00	32.00	40.00	25.00
2.	Potato protection	39.86	59.00	48.00	65.00	39.95	46.00	60.00	30.25	19.00	20.00	06.49
3.	Post-harvest handling	43.13	66.00	53.14	55.00	51.16	55.50	73.00	31.53	28.00	30.00	08.50
4.	Overall	41.83	65.00	55.38	60.00	42.85	50.50	71.00	40.59	26.00	30.00	13.33

Impact of trainings programme on different behavioural aspects of tribal women potato farmers in various aspects of scientific potato cultivation.

As indicated in the enclosed Table 23.16, it could be seen that the percent gain in the knowledge level was highest in the aspect of potato production (65%) followed by post-harvest handling (53.14%) and potato protection (48%). Retention of knowledge was highest in post-harvest handling (51.16%) followed by potato production (41.44%) and potato protection (39.95%). With respect to change in the symbolic adoption and change in the attitude due to training was concerned, again the trends was similar i.e. potato production i.e., 60% and 25%, Post-harvest handling i.e., 31.53% and 8.50% and potato protection i.e., 30.25% and 6.49%, respectively.

Diffusion Effect of the Training Programme

In order to measure the diffusion effect of training on untrained farm women it was necessary to examine quantitatively the segment of the untrained farm women who actually gained knowledge, changes their attitude and symbolic adoption level of improved potato production and post harvest handling practices due to the diffusion effect. For this purpose, the frequency distribution of un trained farm women who witnessed the influence of diffusion was worked out and presented in Table 23.17.

Table 23.17 : Frequency distribution of untrained farm women showing diffusion effect of training (n = 100)

		Diffusion Reported		*Diffusion no Reported*	
S.N.	*Behavioural Dimensions*	*Freq.*	*Percentage*	*Freq.*	*Percentage*
1.	Gain in Knowledge	39	39.00	61	61.00
2.	Gain in symbolic	35	35.00	65	65.00
3.	AdoptionChange in attitude	15	15.00	85	85.00

It is evident from the data presented in the table 20, that 39 percent of the untrained women farmers reported gain of the knowledge, 35 percent reported change in symbolic adoption and 15 percent change in attitude as a result of influence of trained farm women or in other words the diffusion effect of the training programme.

Diffusion effect of training on the various behavioural dimensions of untrained tribal farm women was also ascertained. The findings as contained in the table 23.18 shows that there was about 21 % gain in the knowledge level of untrained women farmers due to the diffusion of knowledge by the trained women farmers. Similarly, percent change in the level of symbolic adoption and the attitude of the untrained women farmers towards improved potato production and post harvest handling practices was 19.5 and 8 percent, respectively due to the diffusion of the information from the trained women farmers.

The reason for the above results could be the fact that interpersonal communication is still a prominent source of communication in the rural system and as result, even the sampled untrained farmers could learn many aspects of the improved potato production and post harvest handling practices from their trained fellow women farmers

Based on the findings, it could be concluded that the training programme had significant impact on all the behavioural dimensions for trained and untrained women farmers. The empirical model emanated from the study has been presented as figure 6.

Table 23.18 : Diffusion effect of training on untrained tribal farm women

		Diffusion Reported		*Diffusion not Reported*	
S.N.	*Behavioural Dimensions*	*Stage of Training*	*Av./Mean Score*	*Mean Differnce*	*% Change in Mean*
1.	Knowledge	Before	40.74	8.62	21.15
		After	49.36		
2.	Symbolic adoption	Before	48.00	9.36	19. 50
		After	57.36		
3.	Attitude	Before	25.00	02.00	08.00
		After	27.00		

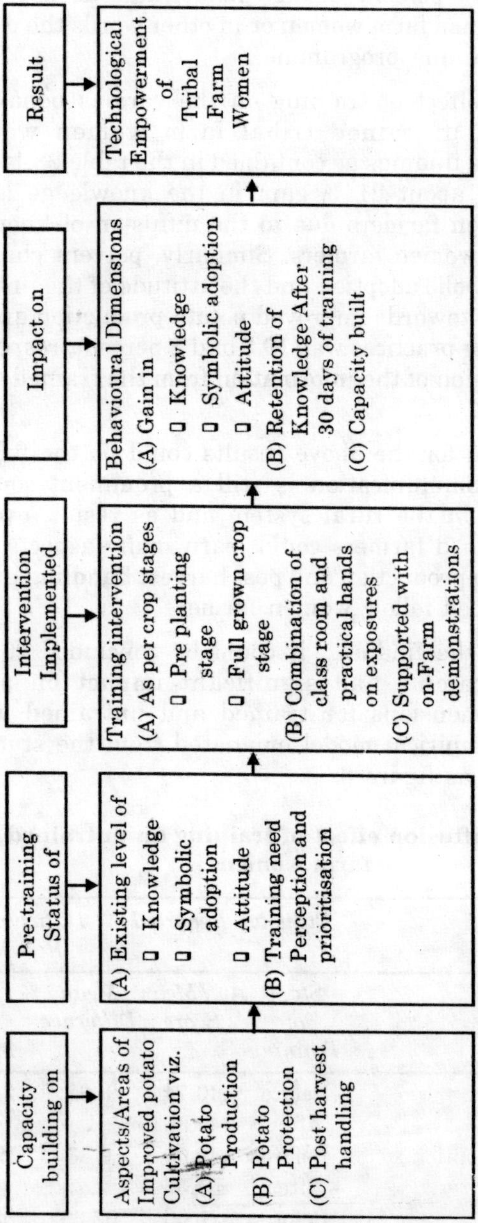

Fig. 2 : Conceptual model of capacity building approach utilized for technological empowerment of tribal women farmers

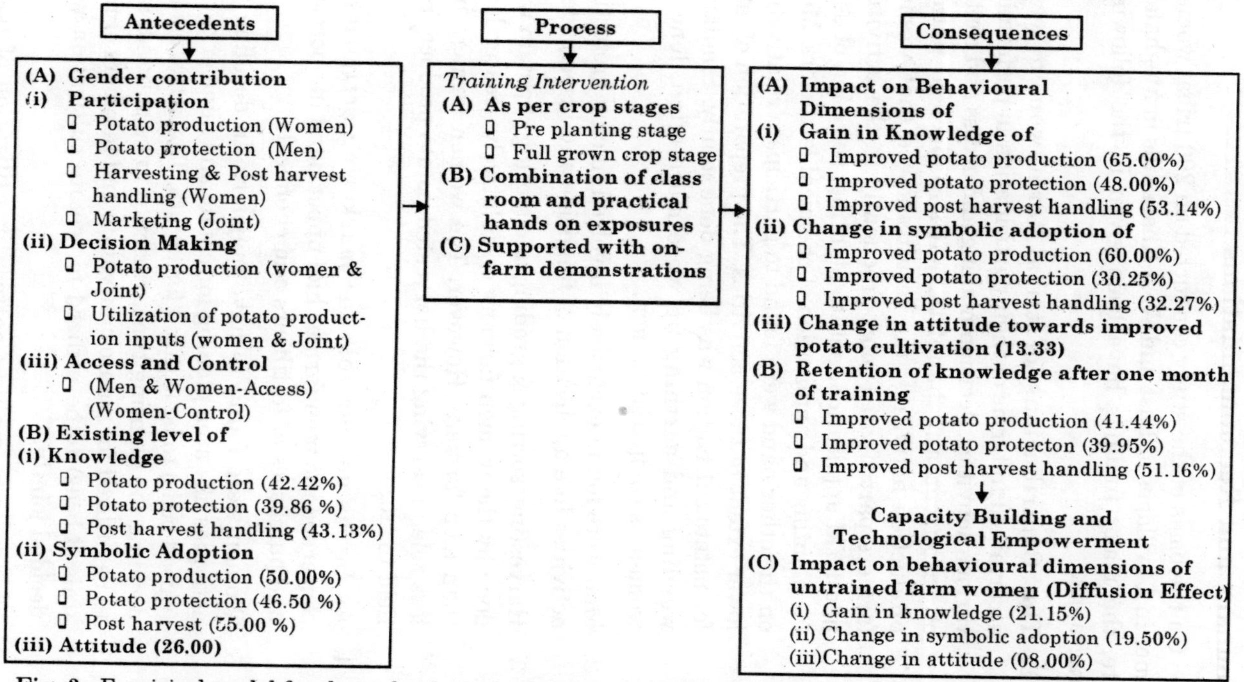

Fig. 3 : Empirical model for the technological empowement of potato growing tribal farm women in Meghalaya

Summary and Recommendations

On the basis of the study conducted on 200 tribal women farmer in 14 villages of East and West Khasi hills in Meghalaya state, the major findings have been given in the following paragraphs.

1. The Contributions of tribal farm women were ascertained in terms of their participation in performing various activities, decision making and access and control of resources and benefits. with respect to potato production activities, the result showed that women farmers participated more frequently in the activities like FYM application, potato planting, selection of seed for planting and application of chemical fertilizers. Men, on the other hand were found to participate relatively a lower extent in these activities. The preparation of land for potato cultivation was being done jointly. Similarly, weeding and earthing up were more often done by women as compared to men.
2. Men were found exclusively involved in crop protection activities like application of fungicide and insecticides.
3. Harvesting, sorting & grading and storing of potato were done by the women farmers on relatively greater level than men farmers. However, The women farmers had less role in carrying the harvested potato produce from field to store.
4. As far as gender role for marketing activities is concerned,the men farmer had upper hand in operation like purchases of fertilizers and chemicals.
5. Women were found to have major rle in deciding the activities like area planing for potato and number of nurs (raised beds) to be utilized for potato. The activities like deciding the planting time, varieties to be planted, land to be utilized for potato cultivation and numbers of nurs (raised beds) to be utilized for potato cultivation were decided jointly.
6. The decision making pattern in utilization of potato production inputs were found similar to the potato

production practices.The other activities like seed source to be utilized, time of application of fungicides, insecticides, etc. and labour to be utilized were predominantly jointly decided except for deciding the amount and sources of other production inputs to be utilized by the women.

7. The study revealed that both men and women farmers had the equal access in use of the production resources like land, labour, and farm implements and seed. However, women had greater control.
8. Access and control over the farm benefits, it could be seen that both men and women had equal extent of access on farm income, seed potato and table potato, but the control over these farm benefits rest mainly with women farmers.
9. On the overall basis, it was found that the tribal women farmers had 42% of knowledge for the improved potato production practices and post-harvest handling. The tribal women had equal knowledge in the post-harvest handling(43%) followed by potato production (42%) and plant protection(39.9%)
10. Overall symbolic adoption level were found about 50 percent. Aspect-wise, the extent of symbolic adoption level were found to be highest in the post harvest handling of potato (55%) followed by potato production (50%) and plant protection (46%).
11. The study also showed that the training need was perceived highest in scientific method of planting (56.7%), followed by balanced use of fertilizer (53.8%) in agronomy of potato. Among the four major areas of plant protection, the training need was perceived highest in the management and control of late blight with need index of 72.1 percent, followed by management of brown rot (63.0 %).
12. The training need for post harvest handling were found to be higher for proper storage of potato (46.3%) while grading & sorting of potatoes the training need index was found to be low i.e. 15.42 percent only.

13. Need index for infrastructural facilities including provision for facilities like cooperative store for availability of fertilizer and agency for timely availability of market information received highest priority by the potato growers in the state which were 66.3 and 59.9 percent, respectively. On the other hand, potato storage related items like facility for and country made storage godowns had low need index i.e. 46.3and 40.0 percent, respectively.
14. A considerable change in the behavioural dimensions of tribal women farmers were also observed. These changes might be due to the training programme given to tribal women farmer. The changes ranged 13.3% in case of attitude to 40.8 %, 42.9% and 55.38 % in change in the symbolic adoption level, retention of knowledge and change in the knowledge level respectively.
15. About 39 percent of the untrained women farmers reported gain in knowledge, 35 percent reported change in symbolic adoption and 15 percent change in attitude as a result of diffusion effect of the training programme. Further analysis revealed that there were 21 % gain in knowledge level, 19.5 % change in symbolic adoption level and 8 % change in the attitude of untrained women farmers due to the diffusion of knowledge by the trained women farmers.

Recommendations

1. The tribal women are the major work force in the potato production as well as storage in the state of Meghalaya. The socio-cultural factors existing in rural Meghalaya allows farm women to have access and control over major production resources and benefits accruing thereon. In addition to, the farm women are the active participants in the decision making with regard to potato production and storage. Therefore, any extension activities/ programmes directed towards potato development in the state need to equally concentrate on the farmwomen along with the men farmers in the state.

2. The approach developed and utilized in the project for technological empowerment of women farmers were found to be highly effective in bringing out the change in the behavioural dimensions of women farmers. Therefore, the devised model on technological empowerment could be further used by the state extension/development functionaries for achieving better results in their similar development endeavours in this hilly and tribal state of Meghalaya.

CHAPTER 25

Potato Storage in the North-Eastern States : Status, Constraints and Potential Interventions

Potato production in the country has reached to record levels in the last two decades and as a result during 2003-04 the total potato production reached to 23.12 million tonnes from an area of 1.27 million ha. The increased production has resulted in several post harvest problems and one of the major problems is that of storage. Potato being semi-perishable crop has to be stored under appropriate conditions most likely in cold stores. However, the inadequate cold storage facilities and its uneven distribution has been a great issue in the country. The other problems associated with potato storage are its bulkiness resulting in transportation difficulties from the farm to stores and adverse environmental conditions during summer months.

The north-east region including seven north-eastern states (Assam, Meghalaya, Arunachal Pradesh, Tripura, Mizoram and Manipur) and Sikkim have lot of topographical variability due to which there is variability in potato crop season as well. Cultivation of potato is done throughout the year in one or the other part of this region starting from plains to high altitudes, and therefore, potato storage has to be done in different ways in different parts and season. Most of these areas being geographically difficult are not well connected for transporting potatoes for storage and above all the cold storage facilities in the region is almost negligible. There are hardly 4-5 cold stores in the region and these too are not being utilized solely for potato storage. Therefore, the cold storage facility is out of reach for the farmer of the region and this results in several problems associated with supply of seed and ware potatoes as mentioned below:

Problems Associated with Storage of Seed Potatoes

(*i*) Farmers of the plain and mid-hill areas are not able to store their produce as seed for the next crop season and hence they have to depend on other states for seed supply, be it West Bengal or farther places like Punjab or Himachal Pradesh.

(*ii*) Farmers of the high hills can store seed potatoes as their harvesting is followed by cool weather conditions, but most of them are unaware of the proper storage practice, which ultimately results in higher storage losses and also undesirable sprouting (i.c.either very long sprouts or unhealthy sprouts).

Problems Associated with Storage of Table Potatoes

(*i*) Due to lack of cold storage facility, the potatoes in the plain areas have to be disposed off quickly after harvesting before they sprout or rot. As a result, the farmers do not get proper price of their produce and compromise with net gains.

(*ii*) The on-farm traditional potato storage methods are not well practiced in plains. This may be either due to lack of knowledge about proper storage practice, resulting in high losses or because of intermittent rains in the area which can sometimes cause failure of on-farm storages.

(*iii*) In the hills, farmers do follow traditional methods of potato storage and the kind of practice varies from place to place. In some places farmers just leave their produce un-harvested (*in-situ*) till they require these potatoes for selling in the market. They even plant the successive crops like paddy in these soils with the idea that the crop will provide shade and cooling effect to these potatoes. Few areas practice storing potatoes in small pits with or without basket enclosures and in other areas farmers just keep potatoes at room temperature by heaping or keeping in baskets, gunny bags or shelves under any sort of shade, wooden structure or hut inside or outside their house (Sah et al, 2006). The structure is generally made of cheap material, which is easily available with the farmers. But here again there are chances for improvement in the present practice so as to reduce the losses.

To sum up the deficiencies associated with potato storage, the biggest problem is that of practically unavailability of cold storage facility for both seed and table potato storage, the others being unimprovised traditional storage methods, lack of knowledge regarding ways to reduce storage losses and incapability or unwillingness to improvise the storage practices being followed at farm level.

Potential intervention to solve the problems of potato storage

The scientists from Central Potato Research Institute, Shimla and its regional stations have studied the problem of storage in the north-eastern region from time to time and have come up with some possible remedies with regards to potato storage which are:

(i) Creation of cold storage facility for seed and table potatoes : The availability of cold storage facility could solve the problem of potato storage to great extent in the region be it for storing seed potatoes or table potatoes. For seed potatoes, storage at 2-4ÚC temperature is a very common practice throughout the country and these kinds of facilities should be created in every state of the region either through state govt, private managers or through co-operatives for proper storage of seed. This will help in saving cost on seed purchase from other states, its transportation and also avoid untimely delay in availability of quality seed.

For table potatoes as well, the cold storage technology is well developed now. Storing potatoes at 10-12ÚC temperature is gaining popularity in the northern states of India and this can very well be adapted in the north-eastern states, once the cold storage facilities are developed in the region.

(ii) On-farm storage for seed potatoes : There is no on-farm method available for storing seed potatoes, which may be used for planting in the next crop season in the plains area where the harvesting is followed by high temperature conditions. Therefore, under such conditions, cold storage is the only alternative. However, in the hilly areas of Meghalaya, Arunachal Pradesh, Mizoram, Nagaland, Manipur and Sikkim, where potato harvesting is followed by low temperature i.e. winter season, the diffused light storage could be used as an alternative.

For the purpose, the potatoes to be used as seed have to be harvested after 10-15 days of haulm killing and after drying in shade for about 15 days for skin curing they are to be treated with 3% boric acid solution for about 30 minutes, dried again in shade and then kept at room temperature under diffused light conditions. The potatoes should be well spread and the layer should not be very thick (only 2-3 potato layers). Spreading of potatoes on the racks made of bamboo or other material on all sides of room will be better. Diffused light will help in slowering growth of sprouts and the sprouts will not grow very long by the next planting season. The low temperature prevailing in the store will keep potatoes better for longer duration and these will be minimum weight loss. This way the seed potatoes may be stored in the hilly areas of the region.

***(iii)* On-farm storage for table potatoes :** Table potatoes may be stored on-farm by traditional methods for a short duration of 2-3 months. There are different kind of on-farm storage methods which are being practiced by farmers since long. Some of the examples include heap storage, pit storage, room-temperature storage, etc. being followed in a large area of Indo-gangetic plains for short duration. By these methods, farmers are able to hold their produce for longer duration and avoid distress sale and also get better price of their produce when the market situations are favourable. However, once the dormancy of potatoes is over, farmers cannot prolong such storages and hence it is suitable only for short duration.

In the north-eastern region as well some traditional potato storage methods are being practiced by the farmers, but they need some improvisations to reduce the losses during storage. In the plains area where the potatoes are harvested between

January to March the heap storage inside rooms or thatched roof may be a good alternative. However, some modifications in the present practice are required as:

- Potatoes should be well cured before heaping.
- Grading and removal of cut, crack and diseased tubers should be done.
- Spreading 2-3 inches thick layer of sand on the bottom of floor may be done before heaping.
- Insertion of a perforated PVC pipe in heap with small lid over the heap helps in little gas exchange with the ambient air.
- Covering of potato heap with a thick layer of straw (1-2 feet) of locally available dried material (rice straw, maize leaves, banana leaves, etc.) helps in maintaining lower temperature of potatoes for longer duration. Presently, the farmers are not following this method.
- Mixing of dried and crushed Lantana leaves with potato helps in reducing the incidence of potato tuber moth. Therefore, in areas where PTM infestation is observed in stores, this may be followed.
- Early heaping before onset of warm weather helps in maintaining lower temperature inside the heaps.
- Once sprouting begins, the potatoes have to be disposed off to reduce losses in term of shrinkage and lower market value.

By taking care of the above points, farmers can store their produce for 2-3 months at their own farm and whenever the market situations seem favourable, they may sell their produce and get better remuneration.

In the hilly areas of the region as well heaping can be done. There is no need for insulation material as the outside low temperature does not allow much weight loss but a thin layer of straw above heaps will help in restricting the entry of light on the tubers and as a result the potatoes will not turn green (greening in table potatoes reduces the market value).

However, the other points like grading, removal of diseased tubers, mixing of Lantana leaves etc. are to be followed for reducing storage losses.

Potato is a very important crop of the north-eastern region and the problems associated with its storage should be given due importance. The policy makers should do the efforts for creation of cold storage facility in these states. The technical know how for improvisation of on-farm potato storage methods is being disseminated by the scientists from time to time and is likely to help farmers in storing their produce in a better way and also to get them better gain from their potato produce.

CHAPTER 26

ON-FARM ACTIVITIES FOR AWARENESS CREATION AMONG POTATO GROWERS

On farm activities like demonstrations, field days, farmers meeting, etc., are the important tools for wider dissemination of potato production, protection and post harvest technologies of potato among the farmers. Therefore, these activities were conducted extensively during the project period. These included management of potato diseases like, late blight, bacterial wilt through chemical control and insect pest like management of potato tuber moth in store by application of locally available lantanna leaves. The impact of these activities was also ascertained. The year-wise status is presented below in Table 25.1 and discussed in the following paragraph.

Table 25.1 : Details of the on-farm activities carried out

		Year-wise Activities Conducted			
S. N.	*Aspects of Plant Protection*	*2003-04*	*2004-05*	*2005-06*	*Technical Observations*
1.	Management of Bacterial wilt by using Stable Bleaching powd	25 nos. of OFTs in Meghalaya	40 nos. of OFTs in Meghalaya	41 nos. of FLDs in Meghalaya	• Incidence reduced to 1-2% from 10-15% • Yield
2.	Management of Late blight in fields	05 nos. of on-farm farmers' meetings	10 nos. of FLDs in Assam	12 nos. of FLDs in Assam	• Incidence reduced by Assam 20-25% Meghalaya : Yield Increrase Assam:10-15%
3.	Management of Potato Tuber Moth	06 field visits to the affected areas in Meghalaya	20 FLDs and 10 Farms meetings in Meghalaya	20 FLDs in Meghalaya	• The damage was reduced to 45% from as high as 80% in control. • Income enhanced by 30-40%

Year 2003-04

1. 25 on-farm trials in East Khasi Hill district of Meghalaya were conducted wherein application of Stable Bleaching Powder was demonstrated for the management of bacterial Wilt.
2. On-farm farmers meeting (5 nos.) were conducted for the management of late blight and other diseases in potato.
3. Field visits (6 nos.) were conducted and potato farmers were persuaded for application ofchopped lantana leaves over potatoes in store for control of PTM

Year 2004-05

(*i*) 10 on-farm demonstrations in three villages in the Barpeta district (Jan 2005) of Assam were conducted for scientific management of Late Blight.

(*ii*) On-farm farmers' meeting **(n = 10)** were conducted for the management of insect pest and post harvest management of potato in Meghalaya.

(*iii*) On-farm demonstrations of application of stable bleaching powder **(n = 40)** for management of bacterial wilt in the potato is in progress in East Khasi Hills district of Meghalaya.

(*iv*) On-farm demonstration (n = 20) on use of lantana leaves were conducted in six villages in the Mylliem circle of East Khasi Hills district in Meghalaya

Year 2005-06

(*i*) 12 on-farm demonstrations in two districts namely Barpeta (6nos in six villages, Jan., 2006) and Kamrup (6 nos. in three villages, Jan., 2006) were conducted for scientific management of Late blight

(*ii*) On-farm demonstration (n = 20) on use of lantana leaves were conducted in six villages in the Myllium circle of East Khasi Hills district in Meghalaya

(*iii*) On-farm demonstrations of application of stable bleaching powder **(n = 40)** for management of bacterial wilt in the potato is in progress in East Khasi Hills district of Meghalaya

The summarized form of results of on-farm activities have been presented below :

Activities done	*Details*	*Year*
1.Assessment of potato varieties for identification of suitable cultivars for different production situations in NE region	For low land production situation in Hill eco system : K. Jyoti (125 q/ha) was assessed as the appropriate variety.For upland and sloppy land production situation in Hill-eco system : K. Girirraj (175 q/ha)For Plains production situation: K. Badsah (150 q/ha) and K. Jyoti (122 q/ha)	2004-05
2. Assessment of TPS hybrids suitable for NE region	92-PT-27 and HPS 1/13 were identified as the best TPS hybrids in NE region on the basis of their seedling tuber yield (120-150 q/ha), shape (oval), flesh colour (yellow), and more resistance to late blight as compared to TPS C3 and other recommended TPS hybrids	2001-04
3.Refinement of farmers traditional method of potato planting in *'Nur-Bun'* method	The seed rate was reduced from 50 q/ha to 33 q/ha, with significantly higher yield (28 q/ha) and better economics (B:C ration = 1:2)	2004-05
4.Refining the late blight management regime in low land production situation	Two spray (one prophylactic spray of Indofil and one spray of Ridomil) helped to give yield advantage of 47% and B :C ratio of 1:2.6 as compared to unsprayed plot without disturbing the prevailing cropping sequence of Potato-Paddy	2004-05
5. Validation of use of organic sources of nutrients for cost reduction and ensuring sustainability of the production situation	Inorganic fertilizers could be safely replaced by bio-fertilizers (*Azotobacter* + *Phosphobacterium* and 10 t/ha FYM) to the extent of 25-30% with comparable yield (24 t/ha)and better economics (1:4.2)	
6.Validation of Management of Bacterial wilt by using Stable Bleaching powder	Incidence reduced to 1-2% from 10-15% and Yield increased by 4-5%	2004-06
7.Validation of use of low cost locally available Lantana leaves for Management of Potato Tuber Moth in country store	The damage was reduced to 4-5% from as high as 80% in control and Income enhanced by 30-40%	2004-06

Impact Assessment of above activities

The above activities led to have following impact at the farmers' field :

(*a*) The incidence of bacterial wilt in low land/lower slope condition reduced significantly to the extent of 10-20% by use of stable bleaching powder. The yield advantage was also noticed to the extent of 15%. Farmers also reported that there was considerably less incidence of other insect pests like cut worm and white grub in the plots treated with the stable bleaching powder as compared to the untreated plot. The secondary impact of this trial helped to increase the demand of SBP among the potato farmers before the state horticultural extension functionaries.

(*b*) One prophylactic spray by mancozeb followed by alternate spray of metalaxyl and mancozeb gave the yield advantage of 15-20% in Kufri Jyoti in Assam condition.

(*c*) The incidence of PTM on the untreated potato in the store was found to the extent of 80% in the East Khasi Hills district of Meghalaya. However, the incidence of PTM after treatment with lantanna leaves was only 4-5%.

(*d*) Several rounds of on-farm farmers' meetings and field days helped to create awareness among the potato farmers of this region and it was also noticed that their knowledge level increased with respect to improved potato production, protection and post harvest handling of potato.

1	2	3	4
May 20-21, 2005	Regional Potato Farm, Mao, Manipur	Director of Horticulture and Soil Conservation, Manipur, Imphal	25, Senapati, Imphal East, Imphal West Thoubal, Bishnupur, Ukhrul.
Dec 9-10,2005	Officers Training Centre, Mantripukhri, Imphal	Director of Horticulture and Soil Conservation, Manipur, Imphal	25, Imphal East, Imphal West, Thoubal, Ukhrul, Chandel, Senapati.
		4. Meghalaya	
Sep 20-21, 2002	District Training Centre, Upper Shillong	Head, Central Potato Research Station, Shillong	Arunachal Pradesh -9; Tawang and West Kameng districts. Meghalaya-23, East & West Khasi hills.Total = 32.
Dec 16-17, 2003	District Training Centre, Upper Shillong	Head, Central Potato Research Station, Shillong	21, East and West Khasi Hills
May 23-24, 2005	District Training Centre, Upper Shillong	Head, Central Potato Research Station, Shillong	27, East and West Khasi Hills
Dec 12-13, 2005	District Training Centre, Upper Shillong	Head, Central Potato Research Station, Shillong	22, East and West Khasi Hills
		5. Mizoram	
March 13, 2003	Directorate of Horticulture, Govt of Mizoram, Aizawl	Director of Horticulture, Govt of Mizoram, Aizawl	25
June 17-18, 2004	Directorate of Horticulture, Govt of Mizoram, Aizawl	Director of Horticulture, Govt of Mizoram, Aizawl	28

1	2	3	4
		6. Nagaland	
Sept 28-29, 2002	ICAR Research Complex for NEH Region, Nagaland Centre, Jharnapani	Joint Director, ICAR, Research Complex, Nagaland Centre, Jharnapani	Manipur-10 Senapati &Imphal districts, Nagaland = 16 Dimapur District, Total 26
Dec 14-15, 2004	IETC, Medziphema	Director Of Agriculture, Govt of Nagaland.	31, Kohima and Phek
May 17-18, 2005	ICAR Research Complex for NEH Region, Nagaland Centre, Jharnapani	Joint Director, ICAR Research Complex, Nagaland Centre, Jharnapani.	27, Peren and Dimapur
Dec, 6-7, 2005	ICAR Research Complex for NEH Region, Nagaland Centre, Jharnapani	Joint Director, ICAR, Research Complex Nagaland Centre, Jharnapani	25, Kohima, Phek, Dimapur
		7. Sikkim	
Sep 16-17, 2002	Directorate of Horticulture, Govt of Sikkim, Krishi Bhavan, Tadong	Director, NRC(O), Pakyong and Director of Horticulture, Govt of Sikkim	30, East, North and South districts.
March 6-7, 2003	Okhrey and Ribdi Villages, . West Sikkim district	Director of Horticulture, Govt of Sikkim	174, West Sikkim
Dec 10-11, 2003	Okhrey and Ribdi Villages, West Sikkim district	Director of Horticulture, Govt of Sikkim	105, West Sikkim
Dec 5-6, 2004	Perbing and Jaobari villages, South Sikkim	Director of Horticulture, Govt of Sikkim	75, South Sikkim

1	2	3	4
Dec 19-20,2005	Directorate of Horticulture, Govt of Sikkim, Krishi Bhavan, Tadong	Director of Horticulture, Govt of Sikkim	40, East, West, South and Nortn all the four Districts
		8. Tripura	
Sept 24-25, 2002	ICAR Research Complex for NEH Region, Tripura Centre Lembucherra	Joint Director, ICAR Research Complex for NEH Region, Tripura Centre	29, All the 4 districts *viz.* Tripura North, South, West & Dhalai
Dec 22-23, 2003	ICAR Research Complex for NEH Region, Tripura Centre Lembucherra	Joint Director, ICAR Research Complex for NEH Region, Tripura Centre	21 from all the 4 districts
Nov 30 Dec 1, 2005	ICAR Research Complex for NEH Region, Tripura Centre Lembucherra	Joint Director, ICAR Research Complex for NEH Region, Tripura Centre	22, Tripura North South and West

Total number of beneficiary farmers : Arunachal Pradesh -74, Assam- 41, Manipur- 117, Meghalaya-93, Mizoram-53, Nagaland-99, Sikkim-424 and Tripura- 72, Grand Total: 973

demonstrations and group discussions. These officers were trained in a way that they could teach the farmers of their area about the improved production technologies of potato in the respective region.

Table 26.2 : Training Programmes for the Officers of NEH

Duration	*Venue*	*Participation*
Sept 2-7, 2002	CPRI, Shimla	Meghalaya-2; Mizoram-2; Nagaland-2 **(Total = 6)**
Dec 14-5, 2003	CPRS, Shillong	Arunachal Pradesh-1; Assam-2; Manipur-3 Meghalaya 5; **(Total = 11)**
May 25-26, 2005	CPRS, Shillong	Arunachal Pradesh 3; Manipur-3; Mizoram-2; Meghalaya-3, Nagaland-2; Sikkim-3; Tripura-2 **(Total = 18)**
Nov 21-25,2005	CPRI, Shimla	Arunachal Pradesh-3; Sikkim-2; Tripura-1**(Total = 6)**
Dec 14-15, 2005	CPRS, Shillong	Arunachal Pradesh-1; Assam-1; Manipur-2 Meghalaya- 7; Nagaland-1; Sikkim-2 Total = 14

Up-gradation of farmers' wisdom about potato cultivation during discourse of each training

Before imparting the training farmers were asked about the improved potato production technologies, suitability of varieties, importance of quality seed and the suitable measures for the control of diseases/pests prevalent in the region. Majority of the farmers were unaware about the improved production technologies, varieties for hilly areas and control measures for diseases/pests. However, farmers were using locally available seed of unknown variety/health status. Farmers used their own practices like heap/pit method and broadcasting of chats at the time of field, preparation/ploughing. This practice is used commonly by the farmers in different states.

During the farmers training in different states the knowledge about improved technologies for production, protection and storage of potato (ware and seed) was provided by resource persons from CPRI. This has led to encouraging response and farmers are now aware of names of potato varieties suitable for their areas, names of the plant protection chemicals, importance of quality seed production technologies (proper spacing, earthing up, etc.). Plant protection measures are now being followed by some of the farmers to save their crop from different diseases/pests, as was told by the scientists in the subsequent trainings.

After the trainings, farmers could know little about the TPS technology that could be used as an alternative for potato production. In the recent training programmes held during May, 2005, the farmers of Meghalaya, Nagaland, Manipur and even Arunachal Pradesh showed greater interest in TPS Technology and almost every farmer demanded for a small quantity of TPS.

During the recent trainings, the farmers wanted quality seed of Kufri Megha which was popular among the farmers due to its stable yield, resistance for late blight and better storability even if it is late maturing. This variety is still grown by many farmers in the East and West Khasi Hills districts of Meghalaya and large areas of Assam.

Besides there is a huge demand for the seed of late blight resistant variety Kufri Giriraj in all the states and Sikkim farmers have a big demand for Kufri Kanchan, the red tuber variety. This is a major impact of the trainings are mentioned as under :

Important Outcome of the Trainings

1. Officers from different states trained by CPRI Scientists are disseminating the knowledge of improved potato production technologies in their areas.
2. Farmers now know about the names of potato varieties suitable for their areas. They are aware about the plant protection measures and methods of application of need based pesticides/fungicides..

3. Use of balanced fertilizer doses.
4. TPS technologies for raising potato crop
5. Sorting, grading of potatoes at harvesting time.
6. Growing ware and seed crops separately.
7. Treatment of seed tubers with non-hazardous chemicals like boric acid.
8. Storage of seed grade potatoes from healthy crop for seed purpose.

Indirect Impact of Training : The demand as well as supply of breeders' seed to these states has increased after the training imparted to the farmers in the region.

Conclusions and Suggestions

Potato is the mainstay of agrarian economy in rural Meghalaya. It is well integrated in the cropping pattern and the dietary habits of the people of Meghalaya. However, the production practices, in general, are traditional in nature. There is an alarming gap in level of adoption of various recommended technologies. The farmers also have many constraints that limit the adoption of recommended production technologies. Thus in spite of its strengths, the potato production in Meghalaya is at a low level. Following interventions are suggested to improve the production levels in the state:

The recommended improved potato varieties for the region have some short comings. Thus earlybulking variety Kufri Jyoti is no more resistant to late blight and Kufri Megha though resistant to blight has less acceptable round tubers and low productivity. The newly introduced variety Kufri Giriraj has high bulking rate, acceptable oval tuber shape and high resistance to late blight disease, but it has poor keeping quality resulting in loss of tubers in fields and stores. There is, therefore, need to develop an early bulking, high yielding late blight resistant variety with oval tubers early or medium maturity and good keeping quality for Meghalaya region.

The traditional potato planting method involves a) high seed rate because of small row to row distance (30-35 cm), and

b) placement of seed tuber, FYM and fertilizer mixture in order reverse to that of recommended method. On-farm trials need to be conducted to demonstrate the relative efficacy of two methods in getting high yields. Possibility of developing a modified method keeping in mind the farmers' land structure and climatic conditions of the region may also be explored.

Meghalaya state is a hot spot for late blight disease. The farmers, however, seldom spray fungicides to protect the crop. They have instead adjusted the planting of summer potato crop to minimize the yield losses. To further reduce the losses, it is suggested that economic fungicide spray schedules for control of late blight may be developed for summer and autumn potato crops, both for upland and low land areas.

Since Meghalaya state has high incidence of brown rot, production of breeder's seed is discouraged. To meet the requirement of potato seed, agro-techniques for production of slightly lower grade of seed need to be standardized and disseminated among potato farmers. Possibility of exploiting TPS technology as a viable means of producing seed may also be looked in to.

A number of traditional low cost potato storage structures and methods are prevalent in the state. There is need to study their efficacy in maintaining quality of seed/table/processing potatoes and if possible suggest modifications for improvement.

For better acceptance of improved potato production technologies, there is need for continued interaction of extension agencies with the farmers. It will also be useful conduct on-farm demonstrations and organize short training programmes.

Assured market price for potato in view of its high cost of production will help in minimizing the farmers' risk in potato production. It is, therefore, suggested that the existing market intelligence system may be improved to provide relevant market information to the potato farmers well in time. This would protect farmers from the distress sale of produce.

PART - 5

EXPERIENCES AND IMPACT OF TRANSFERRING POTATO PRODUCTION TECHOLOGIES IN NORTHERN HILLS

CHAPTER 28

Towards Self-sustaining Cooperative Action through On-farm Potato Adaptive Experimentation

INTRODUCTION

The Farmer Participatory Research (FPR) encompasses a wide array of differing approaches to adaptive type research (Farrington and Martin, 1987; Farrington, 1988). These approaches share the common goal of Incorporating farmers more directly into the process of technology development and application. In general, they focus on agricultural systems involving resource-poor farmers, systems that often have been overlooked by more conventional agricultural research approaches (Chambers *et al.*, 1989). Keeping this in view, *Kat;* cluster of *Mashobra* block of Shimla district of Himachal Pradesh was identified for Lab to Land Programme (LLP) of the Central Potato Research Institute, Shimla. It is situated on Shimla-Chail road, about 32 km away from Shimla at a height of 7300 feet AMSL with a cluster of 12 small hamlets comprising about 130 farm families, all small and marginal. Till 1989-90, it was miles away from scientific know-how on potato as also on other crops The farmers usually grow table potato with less than 90q/ha production by using local seed, planting cut tubers, planting the crop along the slope and hardly applying any fertilizer or plant protection chemicals. However, interesting- point with the farmers was the collective marketing of their produce sending it as far off as Mumbai. Delhi and Calcutta markets. The farmers were keen to improve their production and productivity by the use of scientific way of farming. They were fortunate to have the village *Sarpanch* of progressive outlook as their leader. The farmers took the lead in requesting the Institute to launch LLP in Koti. The programme was, therefore, implemented with an objective to bring the weaker section of society particularly, small and marginal potato growers above the poverty line, and make them self-reliant by generating additional income though their own efforts.

FLOW CHART

FLOW CHART

The details of comibinations of on farm potato adaptive experimentation was as below:

Year/No. of farmers	Particulars		
1989-90 (Adopted Farmers = 60)	➤ On-farm trials (Seed) ➤ Control		
	Components of Technology	**Combination**	
1990-91 (adopted farmers = 60) and 1991-92 (adopted farmers = 70, associate farmers=60)	➤ SEED 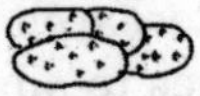	FS + FCP FS + CPRI - T	CPRI-S + FCP CPRI-S & T
	➤ PLANTING METHOD 	FCP FPM + CPRI-T	CPRI - PM + FCP CPRI-PM&T
	➤ LATE BLIGHT MANAGEMENT 	FCP Control, 2 sprays & 4 sprays	CPRI-T Control, 2 sprays & 4 sprays
	➤ FERTILIZER 	FCP	CPRI-RF&T
	➤ HAULM CUTTING CONTROL CPRI-T 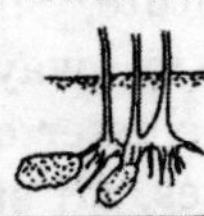		

Where, FS: Farmers seed, FCP: Farmers cultivation practices, CPRI-S&T: CPRI-Seed & Technology, CPRI-T: CPRI-Technology, CPRI-RF&T: CPRI-Recommended fertilizer dose & technology, PM: Planting method, FPM: Farmers Planting method, CPRI-PM&T: CPRI-Planting method & technology

The methodology of the work consisted of the researcher, extensionists and farmers working together for learning by doing and seeing through 'on-farm' potato adaptive experimentation. The emphasis was given on dealing with problems associated with potato-based farming system in the background of formulation of group action and training as the core concept. The opportunity was provided under LLP through self action and technical and critical input support from the sponsoring organisation. Among the adopted and associate farmers group, the key communicators were identified, and then 'stimulated to identify the problems, note relevant information, design and layout trials, evaluate the results and finally mutually share the experiences through regular visits, field days and interaction with key communicators, fellow farmers and the extensionists/scientists.

Each development programme has a philosophy to make the human beings free from traditional bondage of sufferings by making them aware and capable of adjusting and taking of developmental changes and thus be in position to grow in a positive direction systematically and make on their own. Accordingly, the approach explained in the methodology was followed to initiate self and community actualization.

Experiences Gained

The results obtained in the 'on-farm' adaptive experimentation during 1989-90 to 1991-92 are reported in tables 1 and 2 and figure 1.

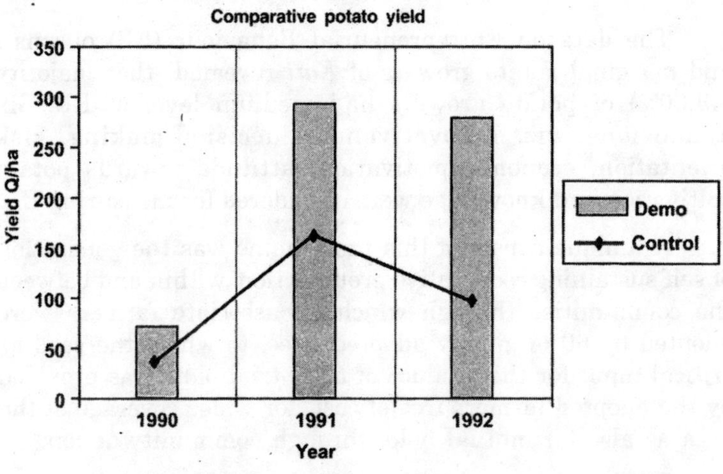

Fig. 1 : Comparative potato yield, LLP, Koti, Shimla

Table : 1 Potato yield of on-farm adaptive experimentation plots (LLP Shimla 1990)

S. No.	*Particulars*	*Yield / Range (q/ha)*	*Time of planting*
1.	Demonstration (N = 60, Seed)	72 (48-80)	22-24 April, 1990
2.	Control (N = 15, Seed)	32 (20-38)	22-24 April, 1990
3.	Table potato (N = 60)	78 (48-110)	3-4 week of March, 1990

The Pre-LLP base line survey results (1989-90) of potato productivity levels obtained with the farmers of *Koti* cluster and marketable surplus quantity showed that 47% farmers of the area could obtain yields ranging between 50-90 q/ha, and only about 60% of the average yield of the state of Himachal Pradesh (115 q/ha). Lack of awareness about the improved technology and soil-climate, disease and pest syndrome, were reported to be the main causal factors.

The data on Entrepreneurial Behaviour (EB) of small and marginal potato growers of *Kati* revealed that majority (43.30%) of potato growers had medium level of EB. Six dimensions, viz. innovativeness, decision-making, risk orientation, economic motivation, attitude towards potato cultivation and knowledge were considered for measuring EB.

A .unique aspect of this programme was the generation of self sustaining cooperative group action within and between the communities through which 60 associate farmers were adopted by 60 originally adopted ones, to whom the seed as critical input for the produce of their trial plots was provided by the adopted farmers free of cost, for wider coverage of the area as also for mutual help, through community action.

The LLP has since been withdrawn from the area, but then the *Kat;* farmers have already reached the take-off stage.

The *Kati* cluster has emerged as quality seed belt harvesting on an average 28t of quality seed potato from a

Table 2 : On-farm potato adaptive experimentation, LLP, Koti, Shimla (1990-91 to 1991-92)

S. No.	*Component*	*Combination + Results (yield : q/ha/range)*					
1.	Seed		FSFCP	CPRI-S+FCP	FS+CPRI-T	CPRI-S&T	
	1991(N=7)		108.7	169.7	156.0	335.5	
			(83.3-140-0)	100.2-222.2)	(118.4-200.0)	(368.5-426.2)	
2.	Planting Method		FCP	CPRI-PM+FCP	FPM+CPRI-T	CPRI-PM&T	
	1991 (N=7)		100.5	120	188.3	289.4	
			(82.2-142.0)	(100.5-150.0)	(175.0-195.8)	(235.2-406.25)	
			102.4	120.4	184.3	284.7	
			(80.0-140.0)	(105.0-140.0)	(170.0-200.0)	(140.0-395.0)	
3.	Late blight management		FCP		CPRI-T		
	(Dithane M-45) sprays	Control	2 sprays	4 sprays	Control	2 sprays	4
	1991 (N=7) 333.1	77.8	103.1	128.1	109.6	266.2	
	(67.5-95.4)	(80.2-130.5)	(112.2-150.0)	(102.0-125.0)	(212.5-318.7)	(262.5-426.2)	
	1992 (N=7) 333.0	77.5	97.8	132.4	109.2	261.2	
	(68.0-95.0)	(70.0-120.0)	(112.0-150.0)	(90.0-125.0)	(212.0-350.0)	(262.5-387.5)	

CHAPTER 29

Entrepreneurial Behaviour of Small and Marginal Potato Growers of Shimla Hills

Intorduction

Himachal Pradesh occupies a unique position in India in that nearly two-thirds of total production of potato in the State is exported out of State to different parts of the country for use as seed. It was thought appropriate to study the entrepreneurial behaviour (EB) particularly of small and marginal potato growers of Shimla hills. Besides, the literature on these dimensions of potato growers is also very scanty. Accordingly, the study-was formulated with the following objectives :

(*i*) Profile the potato growers based on their personal traits;

(*ii*) To study EB of small and marginal potato growers; and

(*iii*) To determine the contribution of different factors associated with EB.

Koti Cluster of Mashobra block of Shimla district identified for Lab to Land Programme of Central Potato Research Institue (CPRI), Shimla in view of the area being predominently a potato belt was the locale for the study. It is situated about 32 km away from shimla at a height of 7300 feet from mean sea level. A total of 60 potato growers adopted under LLP were purposely identified as respondents for the study because of the fact that they were small and marginal farmers.

Variables and their measurements : The EB of the respondents was determined by using the scale developed by Nandapurkar (4) after making the necessary modifications.

Six dimensions viz., innovativeness, decision making, risk orientation, economic motivation, attitude towards potato cultivation and knowledge were considered for the measuring the EB. The EB was taken as the consequent variable of socio personal, economic, potato production and sale and communication-extension variables. Data on these variables were collected with the help of schedule developed for the purpose before the crop season of the year 1990.

Experiences Gained

The Profile of the Potato Growers

The characteristic features of the potato growers are presented in Table 1. The data revealed that more than 76% of the potato growers were more than 37 years of age. Forty percent of the respodnents had studied up to high school level, yet 31% were illeterate. However, this percentage was reduced to 10 as far as education status of the family as a whole was concerned. Majority (60%) of the families had medium to high level of family education status. Thus, despite being remotely located with lack of proper means of communication owing to hilly area, the importance of the education was amply realised by the farm families of the cluster. About 68% farmers had land holding below one ha. As per the agriculture census of Himachal Pradesh also, average size of land holding was 1.30 ha in 1985-86 (2). More than 90% potato growers were keeping animals confirming the observations of Mehta (3) that almost 90% of the farm families residing in the State keep cattle rearing was only a subsidiary activity to agriculture. The animals were reared mainly for draft ad dung which is used mostly as FYM.

'**Table 1 : Profile of the Potato Growers**

Particulars of the Variables	*Frequency (N = 60)*
(I) Socio-personal variables	
(a) *Age*	
(i) Young (36 & below)	14 (23.3%)

(ii) Middle	17
(37-48)	(28.3%)
(iii) Old	29
(49 & above)	(48.4%)
(b) *Self Education*	
(i) Illiterate	19
(34.7%)	
(ii) Primary	12
(20.0%)	
(iii) *High School*	24
(40.0%)	
(iv) Graduate	05
(08.3%)	
(c) *Family Education Status*	
(i) No FES	06
(10.0%)	
(ii) Low	18
(<1.11)	(30.0%)
(iii) Medium to high	36
(1.12-2.44 & 2.45)	(60.0%)
(II) Economic Variables	
(a) *Land Holding*	
(i) Marginal	41
(< 1 ha)	(68.3%)
(ii) Small	16
(1-2 ha)	(31.7%)
(b) *Herd size*	
(i) Small to medium	28
(4 & below)	(46.7%)
(ii) *Large*	28
(5 & above)	(46.7%)
(c) *Orchard*	
(i) Nil	25
(41.7%)	
(ii) Small	15
(iii) Medium	11

(22-51)	(18.3%)
(d) *Annual Income*	
(i) Low	18
(< Rs. 8,000/-)	(30.0%)
(ii) Medium	13
(Rs. 8,000 – Rs. 16,000/-)	(21.7%)
(iii) High	28
(> Rs. 16,000/-)	(48.3%)
(e) *Credit Orientation*	
(i) NIL	14
(23.3%)	
(ii) Medium	10
(< 3)	(16.7%)
(iii) High	34
(> 4)	(56.7%)
(III) Potato Production and sale	
(a) *Credit Orientation*	
(i) Low	16
(< 50.6 q/ha)	(26.7%)
(ii) Medium	25
(50.7 – 90.4 q/ha)	(41.7%)
(iii) High	19
(> 90.5 q/ha)	(31.6%)
(b) *Sale*	
(i) 76.90%	11
(18.3%)	
(ii) > 91 %	41
(68.4%)	
(IV) Communicatgion/Extension variables	
(a) *Use of communication sources*	
(i) Personal Localite	46
* Friends	(76.6%)
* Leaders	11
	(18.3%)
(ii) Personal cosmopolite	14
* VLW	(213.3%)

	* Bank personnel	30 (50.0%)
(iii)	Mass Media	
	*News Paper	17 (28.3%)
	* Radio	31 (51.7%)
	* TV	23 (38.3%)
(b)	*Extension Participation*	
(i)	Low (< 2)	27 (45.0%)
(ii)	Medium (> 3)	26 (48.3%)

Major chunk of (41%) potato growers had no apple orchard. The majority of the potato growers (51.7%) fell in the low to medium income strata. predicitably the high income group were found maintaining apple orchard alongwith the crop production. However 30% of the potato growers were still below poverty line (5 persons with an annual income below Rs. 8,000/-).

More than 72 per cent of the potato growers had medium to high level of credit orientation, underlining that the potato growers of the study area were quite aware about the benefits of utilization of available credit facilities from the cooperative societies and the antionalised banks. Besides, the respondents were also found availing credit for agriculture and dairying which was a healthy indicator of development of both the enterprises.

Productivity level achieved through optimum use of production inputs and quantity of the farm produce earmarked as marketable surplus is of paramount importance to a farm entrepreneur to make farming a commercially viable enterprise. The results peraining to potato productivity levels obtained with the farmers and marketable surplus quantity brevealed that about 42% farmers were able to obtain yields ranging between 50-90 q/ha against the national average of 158 q/ha

(1). Most of the potato growers were selling more than 90% of their potato in market, keeping the rest for domestic consumption and for input as seed, was a healthy entrepreneurial symptom.

Most of the potato growers (76.6%) used fellow farmers as the source of information. Very few were found utilizing the local leaders as the localite source of information. Thus, despite hilly topography, the local leaders were not dominating the scene unlike in remote hilly tribal pockets. As regards personal cosmopolite sources, 50 and 23% potato growers used bank personnel and village level workers as the information sources respectively. Satisfactory trend was observed with regard to the use of mass media. Half of the potato growers reported to have used radio and 38% television as the source of information on agriculture.

The respondents, were found almost evenly divided as having low (45%) and medium (43%) level of extension participation. This indicated that the respondents had fairly participated in different TOT activities in the village and some time outside the village also in a quest to learn to improve their farming.

Entrepreneurial behaviour of Potato Growers

Majority of the potato growers (60%) had medium level of innovativeness (Table 2) Only 13% potato growers had high level of innovativeness. The mean innovativeness index of the respondents was only 18.2% implying that the farmers were though innovative yet had limited scope for being innovative due to ignorance. As regards decision making, 48.3% respondents had below average decision making ability. The respondents average index was 58.7% implying that the potato growers were good in examining the pros and cons and were selecting most efficient means from among the available alternatives for achieving maximum economic returns. However, there is still a scope to improve this ability among the potato growers as almost half of them had below average decision making ability.

The low risk orientation of about 40% potato growers might be atributed to their unsound economic conditions, high probability and severity of risk involved under rainfed conditions coupled with high cost of cultivation and the status quo satisfaction. Almost identical trends was observed in the case of economic motivation as 80% respondents had low to medium level of economic motivation. The mean economic motivation score of the respondents was 66.2% which was considered to he medium. Thus, it could be inferred that on the whole the respondents had moderate economic motivation. The potato growers, therefore, ought to be encouraged to enhance motivational urge for their upliftment.

Table 2 : Entrepreneurial behaviour of the Potato Growers

Dimension	*Category*	*Frequency*	*Mean*
* Innovativeness	Low (< 16.4%)	16 (26.7%)	
	Medium (16.4% - 18.7%)	36 (60.0%)	18.2%
	High (> 18.7%)	08 (13.3%)	
* Decision Making	Below Average (< 58.7%)	29 (48.3%)	
	Average (< 58.8% - 73.7%)	18 (30.0%)	58.7%
	Above Average (> 73.8%)	13 (21.7%)	
* Risk Orientation	Low (<60.0%)	24 (40.1%)	
	Medium (60.1% - 80.0%)	26 (43.3%)	63.7%
	High (> 80.1%)	10 (16.6%)	
* Economic Motivation	Low (< 61.2%)	20 (33.3%)	
	Medium (61.3% - 82.5%)	28 (46.6%)	66.2%
	High (> 82.6%)	12 (20.0%)	
* Attitude	Unfavourable (< 71.3%)	14 (23.3%)	
	Neutral (71.4% - 83.3%)	31 (51.7%)	77.3%
	Favourable (> 83.4%)	15 (25.0%)	
* Knowledge	Low (< 28.3%)	37 (61.7%)	
	Medium (28.4% - 34.2%)	23 (38.3%)	27.5%
Overall EB	Low (< 38.4%)	20 (33.3%)	
	Medium (38.5% - 45.9%)	26 (43.3%)	42.9%
	High (> 46.0%)	14 (23.4%)	

Thus result presented in Table 2 showed that more than half of the respondents had neutral attitude towards potato cultivation. A good number of potato growers had unfavourable attitude towards potato cultivation. The farmers get disillusioned quite often when they do not get back what they had spent on input alone on account of near total crop failure.

The incidence of moisture stress and late blight epidemic were quite high leading to occasional crop failure. Yet, it is highly remunerative crop when saved from vagories of rain fall and diseases. So high risk oriented with positive and neutral attitude farmers need to be motivated and enthused.

Majority of respondents (61.7%) had low level of potato cultivation knowledge. The mean knowledge index was found to be only 27.5%. This is alarming and much needs to be done. Initiating the first line TOt programme under LLP in the cluster needs to be strengthened and extension contacts and farmers' participation in extension programmes need to be boosted.

The EB for all the respondents varied from 30.8 to 59.2% with an average of 42.9%. The potato growers who had high EB were 23.4%. Inspite of the fact that farmers were small and marginal, the trend with regard to EB was found to be encouraging.

Table 3 : Relationship between selected independent variables and entrepreneurial behaviour of respondents.

Independent Variables	*'r' values*
Age (X_1)	0.155
Self Education (X_2)	0.602**
Family Education status (X_3)	0.656**
Land Holding (X_4)	0.001
Herd size (X_5)	0.152
Orchard (X_6)	0.473**
Annual Income (X_7)	0.534**
Use of communication sources (X_8)	0.447**
Extension participation (X_9)	0.477**
Credit orientation (X_{10})	0.232
Yield of potato (q/ha) (X_{11})	0.389**
Sale of potato (X_{12})	0.094

**$p < 0.01$

The zero order correlation presented in Table 3 indicated that 7 out of 12 variables were related significantly and positively with EB at 0.01% level of probability. The socio-personal variables namely, self education and family education status had positive and significant relationship with EB. The economic variables viz., maintaining orchard and annual income showed similar trend like socio-personal variables. These findings are in line with the observations of Nandapurkar (4) who reported that small farmers who had formal education and high income had significantly higher EB. Identical relationship was observed in the case of use of cummunication sources and extension participation. A person utilizing available communication sources and participating in extension activitiesis likely to acquire more entrepreneurial traits which would improve the EB.

Table 4 : Multiple regression between seven independent variables and entrepreneurial behaviour of farmers

Independent Variables	*'b' Values*	*'t' Values*
Self Education (X_2)	0.0275	0.6168
Family Education status (X_3)	0.1420	1.5179
Orchard (X_6)	0.0096	0.5956
Annual Income (X_7)	0.0001	2.5165*
Use of communication sources (X_8)	0.1646	0.9420
Extension participation (X_9)	0.8921	0.6408
Yield of potato (q/ha) (X_{11})	0.0401	2.6469**
	R^2 =	0.6010
	F =	11.0219**

*$p < 0.05$ **$p < 0.01$*

The result presented in Table 4 clearly depicted that yield of potato had positive and highly significant regression coefficient (0.04), whereas, annual income had positive and significant contribution towards dependent variable at 5% level of probability. All the seven variables together could explain a variation of about 60% and the F value was found to be significant at 1% level of probability. Based on these results it could be concluded that yield of potato and annual income were playing crucial role in influencing EB. Thus, to promote EB among the small and marginal potato growers, it is essential to increase the productivity of potato per unit area by employing effective transfer of potato technology with a view to increase their knowledge. These changes may brihg a behavioural change in terms of making the farmers favourable towards potato cultivation which is the key to increase the potato yields. Ultimately this would help in getting more income which is another variable in promoting EB.

CHAPTER 30

IMPACT OF OUTREACH PROGRAMMES IN NORTHERN HILL

Introduction

India has a long history of rural upliftment ageing to almost 9 decades. Non-Government organizations / voluntary agencies were the first to initiate such missions as F. Nicolson's Co-operative Movement (1904), Sevagram (1920), Sriniketan Project (1921), Martamdom Project (1921), Gurgaon Project (1927), Rural Recon-struction Movement (1932), Indian Village Sirine (1931) and Nilokheri Experiri1ent in 1947. During this period and later, Government Institutions also took some initiatives i.e. Grow More Food Compaign (1942), Firka Development Project (1946) and Etawah Pilot Project (1948). These efforts were short lived, limited in geographic cover-age, sporadic and lacked trained manpower, scientific planning and approach as also the technological support. Therefore, there was hardly any impact and professionalism despite a missionary zeal.

In the backdrop of considerable displacement and blood shed on account of partition of the country, prevailing acute rural poverty and hunger, the dream of ushering in *Ram Rajya* through planned social change, the free India's Community Development (CD) Programme launched on October, 2,1952, was the first national level organised effort towards wholistic rural reconstruction. The programme underwent many changes in approach, coverage, scope accommodating additional programmes for weaker sections and under-privileged classes of the rural society and the special programmes launched during sixties, particularly, after green revo-lution. The core programme went through drastic changes in ap-proach and mandate during mid-seventies when the nation shifted to

Training and Visit (T&V) System of Extension.

At present, five extension education systems operate in the country each with its own mandate, approach and thrust. These are:

- First-Line Extension Education System (FLEES)
- National Extension Service (NES) / T&V System.
- Technology Missions on Specific Crops
- Integrated Rural Development Programme and
- Extension efforts by the Non Government Organizations (NGOs)

Whereas, the FLEES, the first system belongs tolndian Cow1Cil of Agricultural Research (ICAR), the second, third and fourth operate under the banner of Ministrie~culture and Co-operation, and Rural Development and the fifth, as the name suggests, operated by theNGOs.

The FLEES of the ICAR is basically an Institutional extension role, a catalytic role, a supportive or complimentary role and a pressure role to accelerate the pace of transfer of technology with minimum distortion/loss, necessiated by the tremendous and complex nature of the technologies generated by the country's research system in mid sixties and onwards. It serves as a window of educational and research Institutions through which the extension agencies as also the farmers can glance through the latest agricultural technologies. The technology transfer programme on potato operated by the Central Potato Research Institute (CPRI), comes under this approach.

TRANSFER OF POTATO TECHNOLOGY IN HILLS

The ToT programme on Potato like the other crops and agricultural and rural enterprises has been in vogue through the Agriculture/Horticulture/Development Departments of the State Governments since the inception of First-Five Year Plan in 1951 and CD Programme as the agency to fulfill the plan objectives. With the advancement of research efforts on crops, more specifically after the Green Revolution in 1966-67, the need for rapid transfer of agricul-tural technology so huge and complex among the ultimate users as production input with minimum distortion and the resultant feed-back for research system to reorient its research programme and provide acid test and judge the adaptability to augment the pace of 'Research-Development-Research Cycle' was felt. This paved the way for initiating FLEES by the ICAR through its Institutes and the State Agricultural Universities ($AUs).

The demands made upon ToT system for accelerating potato production are enormous and formidable, more so in the light of recommendations of the National Commission on Agriculture (1976), that by 2000 AD the area under potato may be increased by 1.5 million ha and yield to 20 t/ha from 16.1 t/ha. Motivated by the track record of potato research achievements and to cope-up with the challenge thrown by the National Commission on Agriculture, the CPR! launched different ToT activities under FLEES, Human Re-source Development Programme (HRDP) and Information, Production and Communication for R&D in Potato during early seventies (Fig. 1).

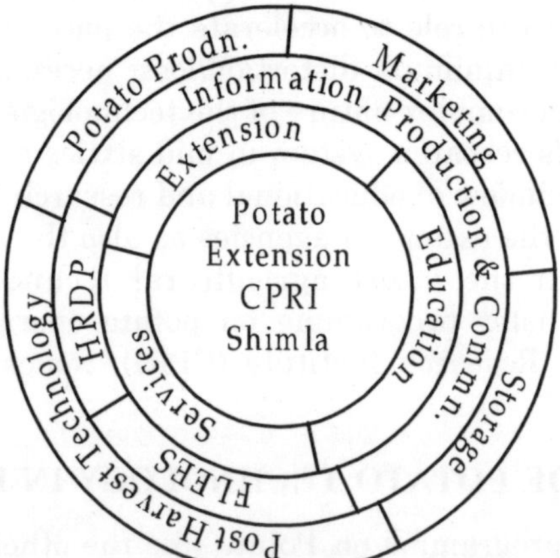

Fig. 1 : Transfer of Potato Technology. **FLEES :** First Line Extension Education System, **HRDP :** Human Resource Development Programme. Information, Production and Communication.

The need for ToT activities was realised in the year 1971 when a publicity section was created in the Institute, which by further strengthening and reorganisation, culminated into a full fledged Divison of Social Sciences. The mandate before the Extension System has been to improve the knowledge of the potato growers, change their skills from traditional to modem, bring about positive change in their attitude towards innovations, create situations congenial to practising such innovations through ensured production inputs and the credit, train a task force of field functionaries for transfer of potato technology, and ultimately, augment production and pro-ductivity. The specific programmes taken up for ToT to meet the above objectives were:

- C perational Research Project (ORP) ~n stepping up seed p ɩtato production
- ORP for tribal area development
- Lab to Land Programme (LLP)
- Trainings for Human Resource Development and
- Information Production and Communication

The focus in all these programmes has been on increased production and productivity of seed and table potato through the use of innovative potato technology by different modes of communication. and training. The programme wise details are spelled out as under:

Operational Research Project (ORP) in Meghalaya

ORP on stepping up seed potato production in Bihar

Is primarily devoted to demonstrating proven innovative potato technologies and testing them for their eco-geoframing system adaptability. The project was started in 1974-75 in Bihar. The clusters of villages were adopted and changed during different phases and various on-farm trials were conducted on farmers fields in selected villages.

The Impact (ORP)

The study into farmers' perception of impact of ORP in Bihar carried out during 1982 -83(4) revealed that 45% of farmers perceived the impact to be of medium nature and 31.6 per cent as high impact (Table 2). Similarly, in 1985-86(1,5) the study on benefit-cost analysis indicated that farmers

Table 1 : Impact of various ToT programmes on potato productivity

Sr. Project and areas of operation	*Year*	*No. of Demonstrations conducted*	*Comparative average yield (q/ha)*		*Percentage Increase in yield*
			Demonstration	*Local*	
(i) (ii)	*(iii)*	*(iv)*	*(v)*	*(v)*	*(vii)*
1. Operational Research Project					
(i) Shillong (Meghalaya)	1976-77 to 1978-79	160	202 (144-252)	174 (142-205)	16.09
2. ORP on Tribal Area Development					
(i) Keylong (Himachal Pradesh)	1981-82 to 1986-87	73	352 (224-506)	283 (225-318)	24.38
3. Lab to Land Programme					
(i) Shimla (Himachal Pradesh)	1979-80 to 1991-92	1100	189 (30-460)	114 (20-162)	85.79
(ii) Darjeeling (West Bengal)	1981-82 to 1982-83	40	132 (126-195)	62 (55-95)	112.90
(iii) Ootacamund (Tamil Nadu)	1981-83 to 1983-84	100	156 (N.A.)	130 (N.A.)	20.00
		2579	219	153	43.1
Overall		5166	253	175	46.67

growing potato with improved methods got gross returns of Rs. 24,250 per hectare as compared to Rs. 7,650 per hectare from traditional methods of potato cultivation. The cost of cultivation with improved methods was Rs. 11,750 per hectare. The major reason for high net returns was high level of productivity (450 q/ha) with improved methods as against 185 q/ha by the traditional methods. Thus, by investing Rs. 100 a farmer would get net returns of Rs. 206 from improved as against Rs. 107 from traditional method.

A probe into knowledge, adoption and constraints analysis of potato technology among ORP farmers of Bhojpur (1988) estab-lished the overall knowledge and adoption indices of farmers to be 70 and 76 respectively (16). The major technological constraints in potato production perceived by the farmers were ignorance about improved cultivars and cultivation practices (49%), time and number of irrigation (36%), weight and size of seed tubers and scientific methods of planting (45%), number of ploughings to be given for field preparation (43%), method, quantity and type of fertilizer for potato (24%), chemicals, their doses and method of application (37%), scientific method of storage of potato (88%), improved culti-vars require more number of irrigation and the farmers did not have assured source of irrigation (69%) and lack of guidance of marketing of potato (29%).

MAGHALAYA

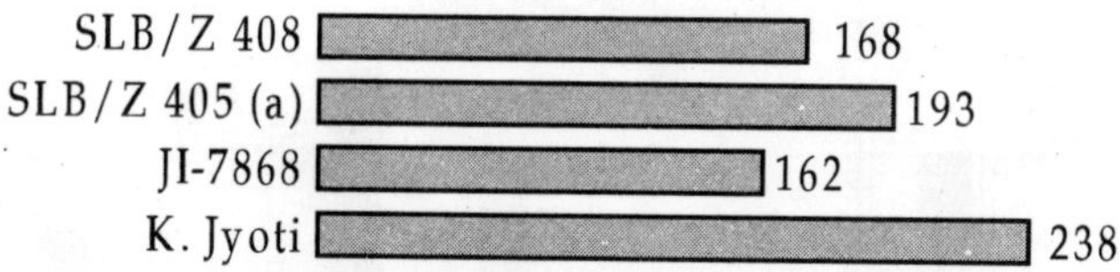

Fig. 2 : Variety wise performance on demonstration plots under ORP during 1974-75 to 1978-79.

Prominent among economic and physical constraints were high cost of seed of improved cultivars and thus not being within the reach of the farmers (94%); improved cultivars require heavy doses fertilizers and fertilizers were costly and hence not within the reach of farmers (94%), potato crop needed more number of irrigations hence very costly (58%), farmers did not get remunerative prices of the produce and so did not favour heavy inve9tments (30%), non-availability of funds to buy costly pesticides (95%), production and productivity level being low and did not call for storage (58%), and land owner did not cooperate on incurring heavy expenditure on land improvement and use of inputs (17%)

Non-availability of seed of desirable cultivar (81%), fertilizer (49%), plant protection chemicals (89%), cold storage facility (98%) at proper time and in close vicinity of the village (81 %) in needed quantity and lack of enough space for storing potatoes scientifically (65%) were the infrastructural constraints inhibiting the farmers in going for scientific cultivation of potatoes.

ORP on stepping up seed potato production in Meghalaya

In Meghalaya, at Shillong, 160 demonstrations were conducted during 1976-79 and an average yield obtained was 202 q/ha as compared to local yield of 174 q/ha (Table 1). Varietal performance revealed that Kufri Jyoti yielded 238 q/ha against 162 q/ha by Kufri Alankar (Fig. 2). The demonstrations on use of mancozeb (Dithane M-45) for late blight control revealed 40 per cent increase in production (Fig. 3).

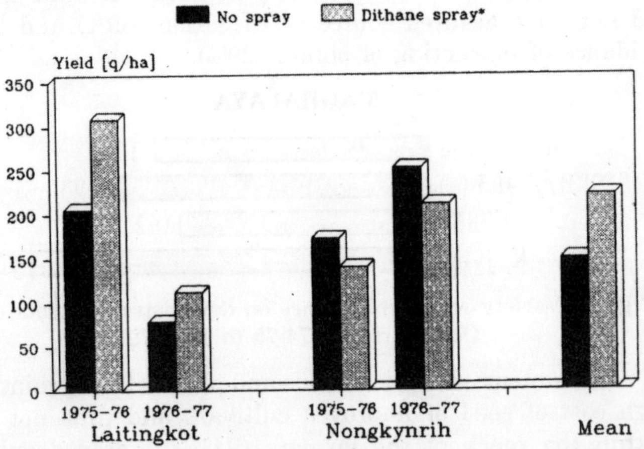

Fig. 3 : Effect of fungicide sprays against late blight on potato yield under OPR, Meghalaya.

ORP for tribal area development

The project meant primarily for the weaker sections of society to ameliorate their living conditions and to bring them in the main-stream of development through additional employment and income generation, was operated during 1980-81 to 1987-88 at three locations, namely; Waghai in the Dangs district of Gujarat, Akole in the Ahmedabad district of Maharashtra.

Spiti district of Himachal Pradesh to promote seed potato produc-tion in these areas. The programme has since been wound up at all the three locations, on fulfilment of objectives, transfer of programme to SAUs or due to limited potential scope for the programme to be operated. The impact of the programme on farmers is discussed below.

At Waghai, the farmers harvested an average yield of 85 q/ha from field demonstrations as compared to local yield of 47 q/ha during 1981 to 1988. The potato was a new crop to the adopted population who generally grew *nagli,* paddy and other small millets. During kharif 1984 (12), a demonstration of comparative economic returns revealed that the farmers got net returns of Rs. 5,242, Rs. 3,296 and Rs. 1,387 per ha from potato, rice and *nagli,* respectively. Although, the cost of cultivation was higher in potato, yet the returns were 1.6 and 3.4 times than rice and *nagli* respectively (Fig. 4).

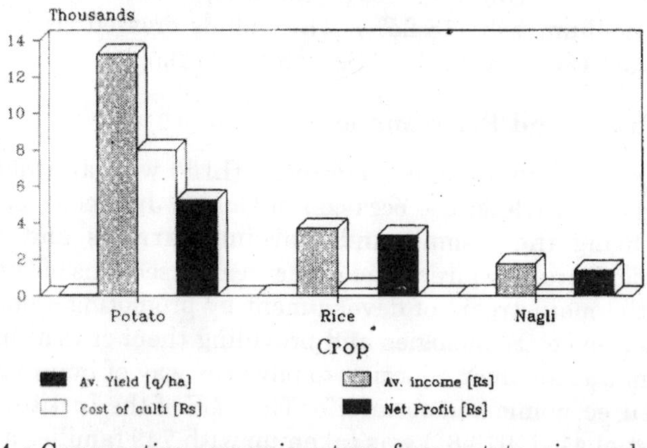

Fig. 4 : Comparative economic return from potato, rice and *nagli* under ORP, TAD, Dangs district, Gujarat, during Kharif, 1984.

Akole, the second location of the project represented the traditional potato growing belt. Besides, conducting yield maximization trials, major emphasis was given to survey the incidence and magnitude of diseases and pests of potato. Early blight and viruses were found to be important diseases in the area with losses ranging from 5-10% and 10-70% respectively. On the basis of 60 demonstra-tions the farmers could obtain yields between 93 to 448 q/ha from on-farm trials as compared to 26 to 159 q/ha from local technology. (Table 1). The project was discontinued in 1984 because of high incidence of diseases and pests and the area could not be converted into a seed pocket.

Lahaul Valley of Himachal Pradesh was the third location. The valley is a major potato belt producing almost one fourth of the total pota to seed produced by HP State. On the basis of 73 demonstrations (1981-87), the potato yields on demonstration plots ranged between 224 and 506 q/ha (Table 1). The incidence of different diseases and pests and varietal mixture were the major problems affecting potato seed health and marketing. The varietal mixture survey during 1987 revealed the overall incidence of mixture of 6.93% in the Lahaul Valley and 11.39% in the Udaipur sub-tehsil. The mixture ranged between 3 to 32%. It was found to be more of Kufri Chandramukhi in Kufri Jyoti. The important diseases in the area were common scab (26-38%), black scurf (9-20%) verticillium wilt (13-25%). The aphids crossed critical limits during the first week of September (14).

Lab to Land Programme

The Lab to Land Programme (LLP) was launched by the ICAR in 1979 on the occasion of Golden Jubilee celebrations, to bring the . small and marginal farmers and landless agricultural labourers and other weaker sections of the society in the mainstream of development by promoting among them the proven technologies and providing them critical inputs as stimulus for their adoption to pave the way of improvement of their economic and social life. The LLP of the Institute, in the first phase (1979-82) was taken up with 749 families at Shimla and other stations of the Institute located in different agro-ecological regions of the country, namely; Shillong (Meghalaya),

Ootacamund (Tamil Nadu) Darjeeling (West Bengal), Jalandhar (Punjab), Patna (Bihar) Modipuram (Uttar Pradesh) and Rajgurunagar (Maharashtra.) In the second phase (1982-84), the programme continued to be operated with 726 families. In the third phase, the programme was confined to only Shimla, Jalandhar, and Modipuram with 526 families. In the fourth phase,t;he programme was implemented at Shimla with 93 families in Matiana cluster and at Modipuram with 59 families in Jaidpur and Ghosipur clusters. In the fifth phase, it was operated with 70 adopted and 60 associate families at Shimla in Koti cluster and 32 families in Kulanjan cluster of Meerut. The on-farm trials matching the number of families adopted were laid,out at different locations.

The average productivity on adopted farmers' fields, on the whole, increased by 78.3, 75.1, 37.1, 12.80 and 98% during I, II, III, IV and V phases respectively on the basis of 2579 demonstrations (Table 1 and Fig.5). More important and significant than productivity level of seed potato on farmers' fields was maxing the farmers self-sufficient with regard to quality seed potato production. At the end of every phase, the selected clusters were not only able to produce the seed fortheir own holdings but also provided to the neighbouring localities. Besides, the programme enabled the farmers to sell their produce at higher rates and thus, generated additional income.

The study on effect of seed tuber quality and fertilizer on the yield of potato conducted at Shimla LLP in 1979-80 (6) revealed that on an average yield level of about 300 q/ha could be obtained using available scientific technology as compared to 100 q/ha with con-ventional methods of farmers of the the area (Table 3.) The returns per rupee invested on fertilizer varied from RS.0.20 to 9.53 with the farmers' seed and Rs. 6.93 to 19.47 with healthy seed supplied by CPR! at different places. SilT).ilarly, on-farm trials conducted in 1989-90 (0) on management of white grub in potato showed that both phorate and quinalphos granules were effective in controlling the white grubs. Higher the elevation in a Valley, lesser was the grub population, perhaps because of low temperature and fewer number of alternative host plants available for the survival of grubs at higher elevation. The cost benefit ratio was 1:6.8 for the quinalphos treated plots and 1:5.6 for the phorate treated ones.

The impact (LLP)

Impact study of LLP phase-I, conducted at Shimla, revealed that the increase in productivity of seed potato among the farmers was 780/0 and the increase in total production of seed potato was 82% when the area under potato increased by 2.1% only. The use of. improved variety increased from 5.8 to 100%, addition of plant protection measures from 6 to 94%, roguing of crop from 24 to 77%, application of fertilizers from 20 to 90%, change of seed after every 3 years from 16 to 100% and proper seedling space from 0 to 50% (1,5, Fig. 6).

Sustainable seed potato cultivation through LLP

Located on Shimla-Chail road, about 34 km away from Shimla, Koti is a cluster of 12 small hamlets comprising about 130 farmers, all small and marginal. Till 1989-90 it was miles away from scientific know-how on potato as also on other crops. The farmers usually grow table potato with less than 90 q/ha production by using local seed, planting cut tubers, planting the crop along the slope and hardly applying any fertilizer or plant protection chemicals. How-ever, a major plus point with the farmers was the collective marketing of their produce sending it as far off as Bombay, Delhi and Calcutta markets. The farmers were keen to improve their production and productivity by the use of scientific way of farming. They were fortunate to have village *Sarpanch* of progressive outlook as their leader. The farmers took the lead in requesting the Institute to launch LLP in Koti. Today, the Koti cluster has emerged as quality seed belt harvesting on an average 28 tonnes of quality seed potato from an hectare (Fig 7). This change came within 3 years of the launching of LLP with the package of scientific know-how transmitted through various media for quality seed potato production.

Each development programme has a philosophy to make the human beings free from traditional bondage of sufferings by making them aware and capable of adjusting and taking advantage of developmental changes and thus be in a position to grow in a positive direction systematically and make on their own. They should be able to set in their own goals of development. Therefore, to initiate self and community actualization, the farmers were contacted individually and in groups. Series of discussion were held on problems of day-to-day concern. A few were identified as community actulizers, key communicators in respect of various problems as well as for further communication of the technology.

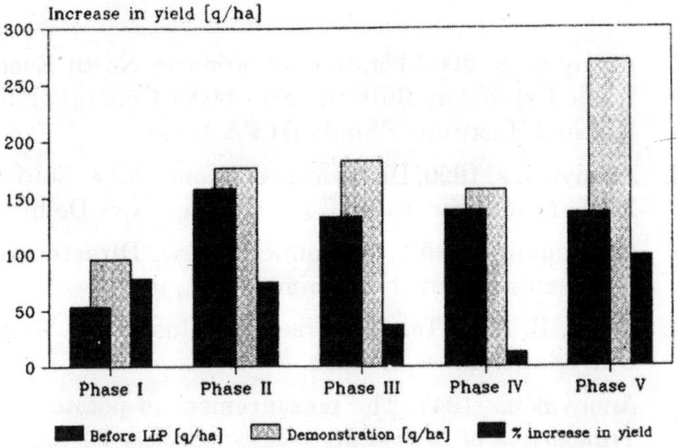

Fig. 5 : Phase-wise progress of LLP in terms of potato yield.

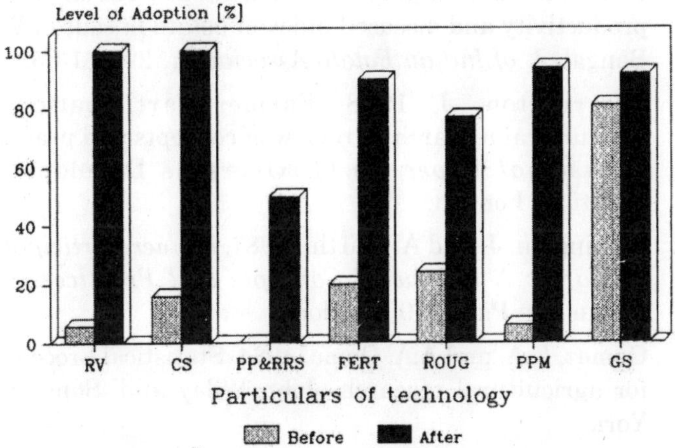

Fig. 6 : Impact of Phase I of LLP, Shimpa,

R.V. : Recommended Variety; **CS :** Change of Seed after every 3 years. **PP & RRS :** Recommended Plant to Plant and Row to Row spacing. **FERT :** Application of Fertilizers; **ROUG :** Roguing, **PPM:** Plant Protection Measures : **GS :** Grading of Seed.

BIBLIOGRAPHY

1. Anonymous. 2003.Potato cultivation in North Eastern India Extension Bulletin No. 34(E) Central Potato Research Institute, Shimla (H.P.), India.
2. Anonymous. 1990, Directorate of Economics & Statistics, Ministry of Agriculture, Govt. of India, New Delhi.
3. Anonymous, 1990, Economic review. Directorate of Economics & Statistics, Shimla, H.P., pp. 16.
4. Abbot, P. 1973. Teach yourself calculus. p 379. English Universities Press, London.
5. Anonymous. 1947. The measurement of potato blight. *Transaction of British Mycological Society* **31**:140-41.
6. Chettri, M, S.S. Mondal and B. Roy. 2002. Influence of Potassium and Sulphur with or without FYM on growth, productivity and diseased index of potato in soils of West Bengal. J. *of Indian Potato Association.* 29 : 61-65.
7. Farrrington, J. 1988. Farmer participation in agricultural research: A review of concepts and practices. *Occasional Paper No.11.* Overseas Development Institute, London.
8. Farrington, J. and A. Martin. 1987. *Fanner Participatory Research:* A *Review of Concepts and Practices. 001,* Discussion Paper 19, London.
9. Gomez, K.A. and A.A. Gome. 1984. Statistical procedures for agricultural research. John Wiley and Sons, New York.
10. Uppadhayay, N.C, Name Singh, Sanjay Rawal and Praveen Kumar. 2003. Response of two potatocultivars to vermicompost and inorganic fertilizers. *J. of Indian Potato Association.* 30(1-2) 85-86.

11. Bharadwaj, K.K.R. and A.C. Gaur. 1970: The effect of humic and folic acids on the growth and efficiency of nitrogen fixation of *Azotobacter chrocochum,* Foliar Microbial. 15: 364-67.
12. Sen, A. and Paul. N.B. 1957. Stabilization of phosphorous by some common bacteria. *Curr. Sci.*26: 222.
13. Sharma, R.C. Sunaina V, Upadhya N.C. and Kumar Mahesh. 1996. Effect of green manuring, organic and bio-fertilizer on tuber production in potato on a typic usto chrept. Proceeding 14th National Symposium on Agronomy environment and food security for 21st Century, CCS, HAU, Hissar, 11-13 December, 1996, 183-84 pp.
14. Singh, K. 1999. Effect of bio-fertilizer and phosphorous levels on the production of Potato *(Solanum tuberosum)* crop under Northeast hill condition. *Indian J. Agri Sci.* **2** :746-49.
15. Bhattacharyya, S.K., S. Phadatare, R. N. Khanna, D. S. Srivastava and B. Prasad. 1993. Efficacy of some fungicides in controlling late blight of potato in India. *Indian Journal of Agricultural Science* **53**:152-57.
16. Bichn, W.L., T. Young, T.R. Dill, J. Snow and E.B. Seifried. 1978. *Phytopath. News* **12**:67
17. Chambers, R., A. Pacey and L.A. Thrupp. 1989. *Farmers first: Fanner innovation and agricultural research,* Intermediate Technology Publications, London.
18. Dutt, B. L., P. N. Mathur and S. Ram. 1982.: Potato in developing countries *Nagaich*, B. B, Ed pp 426 CPRI, Shimla, India.
19. Kadish, D. and Y. Cohen. 1988. Estimation of metalaxyl resistance in *Phytophthora infestans. Phytopathology* **78**:915-19.
20. Kumar Shantanu, U. Sah, V.K. Gupta and C. Deka. 2003. On-farm evaluation of selected hybrid TPS in Meghalaya. *Journal of Indian Potato Association* **30** (1-2): 29-30.

21. Mehta, V.L. 1986, Animal Husbandry Development, Agricultural Research and Development in Himachal Pradesh, Govt. of HP pp. 43-50.

22. Nandapurkar, G.G. 1982, *Small farmers : A Study of their entrepreneurial behaviour.* Metropolitan Book, New Delhi.

23. Schwin, F. J. and P. Morgot. 1991. Control with chemicals. In. Advances in Plant Pathology, pp 225-61.

24. Ingram, D. S. and P.H. Williams Ed.pp 226-61 Academic Press, USA.

25. Singh, D. and P. C. Sharma. 1993. Chemical management of late blight of potato. *Indian Journal of Mycology and Plant Pathology* **24** (2): 143-45.

26. Singh, D., Dig Vijay and J. Pal. 1997. Evaluation of fungicidal spray schedules against late blight (*Phytophthora infestans*) in summer and spring potato in Kangra hills of Himachal Pradesh. *Indian Journal of Agricultural Science* **68** (10): 704-08.

27. Atte, D.O. 1992. Indigenous knowledge as a key to local development: Possibilities, constraints and planning issues in studies in technology and social change No. 20. Ames: Iowa state university, Technology and Social Change Programme.

28. Grenier, L. 1998. Working with Indigenous knowledge- A guide for Researchers, Ottawa: IDRC.

29. Anon. 2002. Basic statistics of Agriculture. Directorate of Agriculture, Assam, Khanapara.

30. Baroh, U. K., Handique P., Baroh S. C.; Barbora, A. C. and Borogohain, R. 2004 Production constraints of rapeseed in Assam. Agril. Extension Review, Mar-April, P:18-20

31. Singh, A. K. (1995). Technological gap in potato cultivation. Agril. Extension Review, Mar-April, pp: 24-25.

32. Gogoi, A. 2002: Production and Marketing of Potato in Jorhat District of Assam. Unpublished M. Sc. (Agri)

Thesis of A.E. F.M, Assam Agricultural University, Jorhat, Assam.

33. Potato Cultivation in North Eastern India, Extension Bulletin N0 34 (E), Central Potato Research Institute, Shimla (H.P),India.
34. Agriculture in Assam, a booklet published by the Directorate of Agriculture, Assam, Khanapara, Guwahati, India
35. Anon. (2000). Basic statistics of North Eastern Region. North Eastern Council, Ministry of Home Affairs, Govt. of India, Shillong.
36. Gaur, P. C. 1996. Manual for True Potato Seed (TPS) Production and utilization. Extension Bulletin No. 30 (E), Central Potato Research Institute, Shimla, (H.P.), India.
37. Gaur, P.C. and S.K. Pandey.1990. Adoption of TPS technology in India. Commercial adoption of True Potato Seed Technology – Prospects and Problems. Central Potato Research Iinstitute, Shimla pp 7-10.
38. Singh, A.N. and Janardan Jee.1990. Economics of potato production from TPS and response of farmers from Bihar. Commercial adoption of True Potato Seed Technology – Prospects and Problems. Central Potato research institute, Shimla.pp 98-106.
39. Sah,U; Kumar,S; Pandey, N.K. and Pandey, S.K.2006.Traditional Potato Cultivation in Meghalaya. Technical Bulletion (72). Central Potato Research Institute, Shimla (H.P.).
40. Praharaj, C.S. Dhruv Kumar and Sharma R.C. 2002. Fertilizer economy in potato production by bio-fertilizers. Potato Global Research and Development, Vol II, pp. 915-919.
41. Anonymous (2002). Basic Statistics of North Eastern Region. North Eastern Council, Ministry of Home Affairs, Govt. of India, Meghalaya.
42. Anonymous (1981). Shifting cultivation in northeastern India. ICAR Research Complex for NEH region, Shillong, Meghalaya.

43. Awasthi, R.P, Kori, S. and Grewal, J.S. (1981). Effect of soil burning on the growth and yield of potato in Khasi Hills, Meghalaya. Indian J. Agric. Sci.51 (5): 312-5

44. Gouroup, P. (1957). Agriculture in Tropics by C.C. Webster and P.N. Webster, pp. 126-150.

45. Griffith, A.L. (1947). Effects of artificial soil regeneration by burning. Indian Forester, P. 526-528

46. Haridson, T.R. (1976). Fire and flame for plant disease control. Annual Review, Phytopathology, pp. 335-379

47. Mishra,V. K. and Gupta, S.K. (2001). Meghalaya Ki Bun Pranali. Kheti. P: 13-15

48. Mukhergee, H. N. (1954). Improvement of soil fertility by application of heat. Proc. Bihar Acad. Agric. Sci. **2**: 1-30

49. Nyle, P. H. and Greenland, D.J (1960). The soil under shifting cultivation. Tech. Commmn No. 51, Commn Bur. Soils, Harpenden, England, P:66-73.

50. Parameter, J. R. and Uhrenhold, B.C. (1975). Suule effect of Pine Needle or Grass smobean. Pengi Phytopathology, Vol. P: 28-31.

51. Singh, A. and Singh, M. D. (1981). Soil erosion hazards in northeastern hill region. Research Bulletin No. 10, ICAR complex for NEH region, Shillong (Meghalaya).

52. Anonymous 2009. Directorate of Economics and Statistics. Department of Agriculture and Cooperation, Ministry of Agriculture, Govt. of India.

53. Gupta, V.K.; Thakur, K.C.; Kumar, Shantanu.; Sah, Uma. 2004.TPS- an alternative technology for potato production in North eastern hill region, Technical bulletin no 64, CPRI, Shimla. P. 2.

54. Kadian, M.S.; Lotha, N.E.; Girish, B.H.; Illangantileke, S.G.; Ortiz, O.; Kumar, Shantanu.; Sah, Uma.; Pandey, S.K. 2010. A Baseline study on potato seed production systems in Meghalaya and Nagaland states of Northeast India. Working paper no 2010-2, Integrated Crop management division, CIP, Lima, Peru.

55. Khatana, V.S., Illangantileke, S.G., ; Upadhya, M.D., Pal, S. and Scott, G.J. 1997. Rapid Appraisal of farm level post harvest handling of potatoes in North east India: The case of Tripura state. Working paper no 11, Social Science Department, International Potato Centre, Lima, Peru, PP. 13.

56. Pandit, Arun; Pandey, N.K.; Rana, R. Kumar, N.R. and Chandran, K.P. 2006. Economic Analysis of Potato Production in Barpeta District of Assam, Indian Journal of Agricultural Economics

57. Upadhya, M.D.; Hardy, B.; Gaur, P.C.; Illangantileke, S.G. 1995. Production and Utilization of TPS in Asia, Proceedings of International workshop held at CPRS, Modipuram, India, Jan 21-30, 1995.

58. Luthra, SK; SK Pandey; BP Singh; GS Kang; SV Singh and PC Pande. 2006. *Potato Breeding in India*. CPRI, Shimla Technical Bulletin No 74, 90p.

59. SK Luthra, SK Kaushik, TA Joseph and Jai Gopal. 2008. Indigenous potato varieties. In "Twenty steps towards hidden treasures-Technologies that triggered potato revolution In India, Eds:SK Pandy & SK Chakrabarty; CPRI, Shimla. p 5-20

60. VK Gupta, Shantanu Kumar, Uma Sah, Shiv Kumar and PH Singh. 2004. CPRI Technical Bulletin No. 62 Quality Seed Potato Production in NEH Region of India pp 33.

61. VK Gupta, KC Thakur, Shantanu Kumar, SK Pandey, and Uma Sah 2004. CPRI Technical Bulletin No. 64 True Potato Seed-An Alternative technology for Potato Production in North-eastern Hill Region pp 22.

62. Pushkarnath. 1976. Varieties and varietal problems. In: *Potato in subtropics*. Orient Longman Ltd., New Delhi. pp 175.

63. Anantharam, M. and Ramanathan, S. (1990). Impact of training on tuber crops. Indian J. Extn. Edn., *26*(1&2): 103 106.

64. Babu, A.R. and Singh, R.P. (1986). Training young farmers towards manpower development in agriculture. Indian J. Extn. Edn., *22*(3&4): 48 53.

65. Chauhan, K.N.K.; Singh, R.N. and Singh, M.P. (1990). Impact of training on knowledge of improved agricultural practices of Arid zone. Indian J. Extn. Edn., *26*(3&4): 64 66.

66. Desai, G. P. and Mohiuddin, A. (1992). Involving women in agriculture: issues and strategies –a study across three states. Journal of Rural Development, (11):5, 637-651.

67. Edwards, A. L. (1969). Techniques for scale construction of attitude. Application Century crafts

68. English, H. B. and English, A. C. (1961). A comprehensive dictionary of psychological analytical terms. Green & Company, Longmans, New York.

69. Fulzele, R.M. (1986). Multidimensional analysis of training programme of Krishi Vigyan Kendra. Unpublished Ph.D. Thesis (NDRI), Kurukshetra University, Kurukshetra.

70. Fulzele,R.M; chand,Ram; mali,B.S and Chauhan,J.P.S (1995). A study of differntial impact of KVK training programme on dairy farmers of different states. Staff Research Council Project Report, NDRI, Karnal.

71. Gour, R.A. and Shrivastava, K.S. (1983). Analytical study of the training programme of farmers' training centre, Jabalpur (M.P.). Maha. J. Extn. Edn., *11*(2): 53 56.

72. Johnson,H.Sam (1987). Agricultural Research Policy and development. FAO (UN), Ed. Vernon.W. Ruttan.

73. Kamalsen, P.S. (1971). A study on the effectiveness of one-day farmer training camp under high yielding varieties programme in Trivendrum district, Kerala. Unpublished M.Sc. Thesis, Tamil Nadu Agril. Univ., Coimbatore.

74. Klongan, G. E. and Coward, E. W. (1970). The concept of symbolic adoption: Suggested interpretation, Rural Soc., 35(1):73-83

75. Perumal, G.; Pomusamy, K.A. and Sujatha, J. (1993). Alternative model for T&V system - An action research. Indian J. Extn. Edn., *29*(1&2): 1 9.

76. Reddy,P.R (1989). Impact of farmers training programme on knowledge level and adoption behaviour of trained and untrained farmers. Maha. J.Extn.Edn., 8: 85-90.

77. Rogers, E.M. and Shoemaker, F.F. (1971). Communication of innovations. Collier MacMillan Publishers, London.

78. Sah, U. (1996). Impact of scientific dairy farming training programme of KVK, NDRI, Karnal on farm women. Unpublished M.Sc. Thesis, NDRI Deemed University, Karnal.

79. Satyanarayan, M.; Husain, S.M. and Reddy, P.R. (1994). Effectiveness of institutional training programme for tribal programme. Maha. J. Extn. Edn., *13*: 281 283.

80. Singh, M. and Singh, M.P. (1990). Training affecting symbolic adoption: A multivariate approach. Indian J. Extn. Edn., *26*(1&2): 61 66.

81. Singh, S.S.P. (1968). An analysis of farmers' training programme in Bihar. Unpublished Report, Bihar Agril. College, Sabour.

82. Swaminathan, M.P. (1989). Impact of training programme in tea nursery management in Nilgiries, UPASI KVK Coonoor, *32*(4): 15 20.

83. Umamahesha, A.B. and Channegowda, M.B. (1988). Impact of peripatetic training on attitude of farmers. Indian J. Extn. Edn., *24*(3&4): 43 49.

84. Vashistha, S.B. (1987). Farmers' training for agricultural development in India. Deep and Deep Publications, New Delhi.

85. Venkatesan, D. (1993). Strategies for agricultural development in tribal sub plane area. Yojana, *37*(20): 13 15.

86. Kumar, S; Singh, P.H., Gupta, P.H.; Sah, U and Pandey S.K. (2006) Integrated development of Horticulture in North Eastern states of India (MM-I), Technical bulletin no 76, Central Potato Research Institute, Shimla.pp 4

87. Gupta, V.K.; Thakur,K.C.; Kumar,S.; Pandey,S.K. and Sah,U.(2004). True Potato Seed.

Annexure 2

Optimum planting and harvesting time in different part of NE region

Height above Sea level	*Planting* time	*Harvesting* time	*Part of NEH* region	*Remarks*
Plains (up to 500 m)	Sept.-Oct.	Jan-Feb	Assam, Tripura, Manipur valley, Dimapur district of Nagaland and and lower parts of Arunachal Pradesh.	Irrigation is necessary for good crop growth.
Mid-hills (up to 1500 m)	Feb. - March	End June to end July	Siang district (Basar area) of Arunachal Pradesh, Jowai subdivision of Jaintia hills, some areas in Khasi hills.	Rainfed crop
High hills (up to 2400 m)	Mar. - April	July - August	Parts of Arunachal Pradesh, Upper Shillong, Mawphlang and Mairang area of Meghalaya, Mao and Ukhrul area of Manipur, Blue Mountains in Mizoram	Rainfed crop
Very high hills (above 2400 m)	-do-	-do-	Tawang, Se la upper Subansiri, Siang and Kameng districts of Arunachal Pradesh.	Rainfed crop

INDEX

R

S

T

U

V

W

Y